Healthcare Engineering

Rajeev Shorey · Purnendu Ghosh
Editors

Healthcare Engineering

Proceedings of CAETS 2015 Convocation on Pathways to Sustainability

Springer

Editors
Rajeev Shorey
Indian National Academy of Engineering
Gurgaon, Haryana
India

Purnendu Ghosh
Birla Institute of Scientific Research
Jaipur, Rajasthan
India

ISBN 978-981-10-3110-6 ISBN 978-981-10-3111-3 (eBook)
DOI 10.1007/978-981-10-3111-3

Library of Congress Control Number: 2016956185

Printed on acid-free paper

This Springer imprint is published by Springer Nature
The registered company is Springer Nature Singapore Pte Ltd.
The registered company address is: 152 Beach Road, #22-06/08 Gateway East, Singapore 189721, Singapore

Foreword

I am delighted that the Post-Proceedings of the International Council of Academies of Engineering and Technological Sciences (CAETS) 2015 Convocation on "**Pathways to Sustainability: Energy, Mobility and Healthcare Engineering**" held in New Delhi, India, on 13 October–14 October 2015, are being brought out. CAETS, being a non-governmental international organization of Academies of Engineering from 26 countries, which has the broad objective of advising governments and providing an international form for addressing all issues of concern pertaining to engineering and technology, plays a vital role in the promotion of the engineering profession. Indian National Academy of Engineering (INAE), a Member Academy of CAETS, is proud to have had the opportunity of hosting the CAETS 2015 Annual Meeting and Convocation at New Delhi, India.

The CAETS event brought together the eminent engineering luminaries and bright young engineers from 24 countries on a single platform who shared their country-specific knowledge on the themes of the convocation. The CAETS 2015 Convocation focussed on the topics of Energy, Mobility and Healthcare Engineering which are of topical interest all over the world in recent times. Energy security, provision of economical modes of transport and mobility, and affordable and high-quality facilities for health care are necessary for the citizens of the world in order to lead comfortable and secure lives. It is indeed apt that the convocation addressed key issues in the fields of Energy, Mobility and Healthcare Engineering that play a vital role in the lives of mankind.

It is a matter of pride that the convocation had representation of eminent engineers from academia, R&D and industry from over 24 countries across the globe who shared their valuable experience and expertise and enriched the deliberations with their thought-provoking presentations. These presentations were indeed brilliant and stimulating and have resulted in interesting papers which are showcased in the Post-Proceedings. I thank all the authors for devoting their precious time for contributing papers for the proceedings.

I express my sincere thanks to the editors of the CAETS 2015 Convocation Post-Proceedings for their untiring efforts at every stage of planning of the event and in publication of the Post-Proceedings. The assistance and initiative taken by Springer in bringing out the publication are deeply appreciated.

On behalf of Indian National Academy of Engineering (INAE), it is my proud privilege to express my deepest gratitude to the CAETS organization for giving us an opportunity to host this event of international repute for the first time in India.

I sincerely hope that the event left many pleasant and memorable memories for all the delegates, and I am certain that the Post-Proceedings of the CAETS 2015 Convocation on "Pathways to Sustainability: Energy, Mobility and Healthcare Engineering" shall be well accepted in the engineering realm and by the engineering community.

New Delhi, India

Dr. B.N. Suresh
President, INAE

Preface

Preamble

The CAETS CONVOCATION-2015, hosted by the Indian National Academy of Engineering (INAE), was held on 13 October–14 October 2015 at New Delhi, India. The CAETS Governing Council had chosen the theme on sustainable engineering pathways for energy, mobility and healthcare sectors for the conference attended by 250+ delegates comprising of the Fellowship of CAETS member academies, senior engineering experts, scientists, thought leaders, policy makers and industry executives from Belgium, Canada, China, France, Germany, Hungary, India, Japan, the Netherlands, South Africa, Spain, Sweden, Switzerland, UK, USA, Uruguay and other countries. This international event was formally inaugurated by Dr. Harsh Vardhan, Hon'ble Minister of Science and Technology, Government of India, and he assured that the recommendations of the CAETS Convocation will be acted upon by the government. Professor Asutosh Sharma, Secretary, Department of Science and Technology, Government of India, addressed the gathering and highlighted the importance of the selected themes to developing nations.

The two unique features of CAETS-2015 were the special gesture made by the INAE (a) to provide 50 % international travel support and complimentary accommodation to encourage young engineers below 45 years of age to deliver lectures in the technical sessions along with eminent senior engineers and (b) to organize poster sessions and awards to encourage very young engineers and research scholars from CAETS member academies to present their research work. These initiatives paid rich dividends by attracting the participation of 30+ young engineers in CAETS Convocation. Another major initiative is the formal launching of the electronic version of the CAETS report on "Transitioning to Lower Carbon Economy: Technological and Engineering Considerations for Building and Transportation Sectors" during the inaugural function of CAETS CONVOCATION-2015 by the chief guest. This is a joint initiative of 12 CAETS Member Academies from Australia, Canada, China, Germany, India, Japan, Korea, South Africa, Switzerland and UK. The experts from these academies met at Capetown (South Africa), Beijing

(China), Berlin (Germany) and London (UK) during 2013–2015 to identify the technological and engineering priorities and emerging options for transition to lower carbon economy in building and transportation sectors.

Recommendations

Energy

Global energy consumption will grow from 524 to 820 quadrillion BTU between 2010 and 2040, leading to GHG emission increase by more than 40 % from the present level under business as usual scenario. Since the dynamics of energy growth is driven by the population and wealth enhancements, it is important to keep track of these growth dynamics in different countries. Energy resources sustainability as well as their cost efficiencies would dictate as rapid shift as possible to lower carbon non-fossil energy systems. Since under this context, recent developments in three-stage thorium-based nuclear energy, large-capacity solar thermal power plants, fuel cell-based independent propulsion systems for strategic and civilian applications, wind-generated hydrogen as a renewable energy carrier and unconventional gas from coal seams, underground coal gasifications and biohydrogen from waste resources provide potentially lower carbon energy options. A joint voluntary initiative by CAETS member academies from Argentina, Australia, Canada, China, Germany, South Africa, Switzerland and UK on unconventional gas involving basic and social scientists, engineers and economists on unconventional gas provides a new model for CAETS member academies to emulate in the future.

Developing energy scenarios under various application modes is a powerful mechanism for devising future energy-efficient systems. A typical example is the highly interconnected electricity, transport and heating sectors. The coal-based energy continues to attract engineers to develop the cleaner process options and to minimize the GHG emissions keeping in mind the transitional priorities. The actions include upgradation of beneficiation, ultra-supercritical combustion and integrated gas combined cycle concept for gasification, cofiring of coal and biomass, and China's achievements in some of these areas are noteworthy. In energy materials, in which significant developments are taking place worldwide, we have to look beyond silicon for next-generation solar cells by adopting conjugate organics, inorganic quantum dots and mixed semiconductor oxides/peroxides. Similar challenges are foreseen in high-energy-density battery and membrane materials for fuel cells. Light-based technologies can play a major part in developing future energy-efficient systems. There is a need to integrate photonic and biophotonic technologies. In this international year of light technologies, such initiatives have to be taken to achieve multi-functionality, tunability, low power for fast operations, reconfigurability and cost-effectiveness so much necessary for energy, mobility and healthcare sectors.

The transition to lower carbon regime requires a careful assessment of current energy and emission loads at national level for setting energy/emission reduction targets and timelines. The selection of cost-effective technology and engineering options and their incentivization for creating new markets are the major challenges. The expert panel, which deliberated on these issues, stressed the need to develop high-strength lightweight materials including nanocomposites and nanostructured steels and bio-, electrochemical and catalytic options for hydrogen generation adaptable for building and transportation sectors. CAETS need to work closely with related international bodies such as World Economic Forum, International Panel on Climate Change and allied agencies.

Mobility

The future engineering challenges lie in the development of semi- and fully automated transport vehicles, intelligent driver–vehicle interface and innovative multi-axle hydraulic trailers for heavy loads. The fast-changing digital technology has opened up new avenues for electrifications of rail, air and marine transport vehicles. The five digital forces, viz. cloud computing, mobile technologies, social networks, big data and robotics, will make high impact on these developments.

Exciting developments are taking place in bridge design and construction technologies with urban and rural transport restructuring, virtual mobility and carbon footprint minimization during their construction as defining factors. Novel retrofit technologies are needed for their restoration. A life cycle approach embracing bridge design, construction, maintenance, restoration and dismantling is a future priority area for civil engineers.

The recent developments in rapid urban rail transportation systems in China, India and Japan have demonstrated the new engineering skills acquired in the planning, design and execution of underground tunnels and structures. From material engineering point of view, material recycle and functionalized material applications are high priority areas in evolving sustainable roadways.

The dilemmas in mass transportation in emerging economies are many. It is important to consider seamless connectivity, smart mobility, enforcement of advanced safety measures, equitable allocations of road space for multiple type of vehicles and time-variant traffic demands.

Health Care

Multitude of engineering challenges await healthcare sector in terms of new diagnostic tools, next-generation medical devices and application of informatics and analytics. The recent advances in nano- and point-of-care diagnostic, scalable medical and remote neonatal monitoring systems have enhanced the chances of early and accessible diagnosis. Structural process concept has greatly helped in developing groundbreaking technologies for these systems in advanced nations with the help of multi-disciplinary teams comprising engineers, product designers, business analysts and clinicians.

The concept of regenerative engineering which combines tissue engineering, material science, cell physics and developmental biology has enabled the technologies for next-generation medical devices. Groundbreaking bionic ear and eye technologies have enabled notable progress against hearing and visual impairments. The recent advances in sensors, telecommunication and mobility engineering will play a major role in evolving next-generation device technologies for both communicable and non-communicable diseases. Sensing and data analytic skills provide new transformational material opportunities in intensive health care. Advanced computer-aided tools based on big data analytics are needed for biologically meaningful insights into the enormous volume of microbiome data generated from the sequencing platforms.

A panel of experts examined the issue of convergence of engineering and healthcare sciences. They are certain that the convergence is occurring impressively due to the application of ICT and big analytics in healthcare systems. Several key issues including affordability of well-engineered systems in rural and urban environments and engineers role in new drug discovery came up for discussion.

The major takeaways from CAETS Convocation are many. Achieving long-term sustainability under business as usual mode is virtually impossible in energy, mobility and healthcare sectors. Major emphasis has to be placed on energy consideration, expanded use of non-fossil primary energy, decarbonization of existing energy sources and enhancing the energy efficiency of individual systems. Higher investment in R&D and demonstration is essential for technologies which are close to market maturity and those requiring scale-up. Commitment from policy makers, different stakeholders, academic community and market leaders is very essential for the commercial realization. Ethical practices in engineering are vital for achieving high level of successes in every application field particularly so in creating human artefacts that are non-existent in nature. Gender enhancement in engineering design endeavours is essential for developing rational systems. Younger engineers have to be given increased responsibilities in evolving innovative engineering systems. The demand for novel engineering solutions will increase exponentially while developing smart energy, mobility and healthcare systems. Research, prototyping and technology transfers in such systems have to be pathbreaking in nature. The CAETS Engineering Community is committed to create a conducive environment for open access information sharing in the above sectors between the member academies.

Gurgaon, India Rajeev Shorey
Jaipur, India Purnendu Ghosh

About the Book

The book contains the proceedings of CAETS 2015 Convocation on "Pathways to Sustainability: Energy, Mobility and Healthcare Engineering" that was held on 13 October–14 October 2015 in New Delhi. This 3 volume proceedings provide an international forum for discussion and communication of engineering and technological issues of common concern. This volume talks about 'health care' and includes 11 chapters on diverse topics such as regenerative engineering, big data analytics in health care, molecular science, rising expenditure on health issues and adoption of personalized medicine. The contents of this volume will be useful to researchers and healthcare professionals.

Contents

About the Editors

Dr. Rajeev Shorey is the Principal Scientist at the TCS Innovation Lab, Cincinnati, USA, and Bangalore, India. Dr. Shorey received his Ph.D. and MS in Electrical Communication Engineering from the Indian Institute of Science (IISc), Bangalore, India, in 1997 and 1991, respectively. He received his B.E. degree in Computer Science and Engineering from IISc, Bangalore, in 1987, and the B.Sc. degree from St. Stephen's College, Delhi University in 1984. Dr. Shorey's career spans several reputed research laboratories—General Motors (GM) India Science Laboratory (ISL), IBM India Research Laboratory and SASKEN Technologies. He was an adjunct faculty in the Computer Science Department at IIT, Delhi, from 1998 to 2005. He was a faculty in the Computer Science Department at the National University of Singapore from 2003 to 2004, while on leave from IBM Research Labsoratories in New Delhi.

Dr. Shorey's work has resulted in more than 50 publications in international journals and conferences and several US patents, all in the area of wireless and wired networks, including wireless security. He has 12 issued US patents and several pending US patents to his credit. His areas of interest are wireless networks, Internet, telecommunications, telematics, data security and data analytics. Dr. Shorey has served on the editorial boards of IEEE Transactions on Mobile Computing and is currently serving on the editorial board of WINET (Wireless Networks Journal of Mobile Communication, Computation and Information) journal. He is the editor of the book titled "Mobile, Wireless and Sensor Networks: Technology, Applications and Future Directions" published by John Wiley, USA, in March 2006. Dr. Shorey has given numerous talks, tutorials and seminars in industry and academia all over the world. He is the founding member of the Communication Systems & Networks (COMSNETS) conference in India.

For his contributions in the area of communication networks, Dr. Shorey was elected as Fellow of the Indian National Academy of Engineering in 2007. Dr. Shorey was recognized by ACM as a Distinguished Scientist in December 2014. He is a Fellow of the Institution of Electronics and Telecommunication Engineers, India.

Purnendu Ghosh is the executive director, Birla Institute of Scientific Research, Jaipur. Prior to this, he was professor at Indian Institute of Technology, Delhi. He worked at the Swiss Federal Institute of Technology (ETH), Zurich, Switzerland; University of Melbourne, Australia; and National Research Centre for Biotechnology (GBF), Braunschweig, Germany. He has taken a lead role in augmenting bioprocess engineering and biotechnology activities in the country as a member of several Task Forces of the Department of Biotechnology, Government of India. He is currently the chief editor of Publications of Indian National Academy of Engineering. He is a columnist of a leading newspaper group. Besides technical publications, he has written books of different genre.

Ethics-Based-Engineering: Importance of Academy's Initiative Toward Human Security and Well-Being

Hideaki Koizumi

Abstract Marie Curie (1867–1934) and Pierre Curie (1859–1906) received the Nobel Prize in Physics in 1903 for the discovery of polonium and radium, which eventually led to nuclear energy. At the award ceremony, their lectures concluded with Pierre's remarks to the effect that "science is neutral in values. Whether to use the results of scientific research for good or bad purposes is entirely up to the humans who use them. Are humans really mature enough to use science only for good purposes? I hope we are."

Keywords Engineering ethics · Warfare technology · Human ethics · Human security · Ethics for sustainability

1 Introduction

Marie Curie (1867–1934) and Pierre Curie (1859–1906) received the Nobel Prize in Physics in 1903 for the discovery of polonium and radium, which eventually led to nuclear energy. At the award ceremony, their lectures concluded with Pierre's remarks to the effect that "science is neutral in values. Whether to use the results of scientific research for good or bad purposes is entirely up to the humans who use them. Are humans really mature enough to use science only for good purposes? I hope we are."

A paper for the CAETS (International Council of Academies of Engineering and Technological Sciences) Convocation to be held in New Delhi, India, October 12–16.

H. Koizumi (✉)
Engineering Academy of Japan (EAJ), Tokyo, Japan
e-mail: tanaka@eaj.or.jp

H. Koizumi
Science Council of Japan (SCJ), Cabinet Office of Japan, Tokyo, Japan

H. Koizumi
Hitachi, Ltd., Tokyo, Japan

R. Shorey and P. Ghosh (eds.), *Healthcare Engineering*,
DOI 10.1007/978-981-10-3111-3_1

As is widely known, the Nobel Prize was initiated in accordance with the dying words of Alfred B. Nobel (1833–1896), who wanted to recognize the development of scientific technologies and peace using the enormous assets acquired from his invention of dynamite as the original funding. Dynamite helped save many people's lives by enabling large-scale flood control works, but it also was used as a tool of massacre in warfare, becoming a typical example of a scientific technology with a double-edged sword, reflecting the bright and dark sides of science.

The case of Fritz Haber (1868–1934) led to further consideration of the importance of the ethics of scientists and engineers. Haber showed significant talent in physics, chemistry and biology and developed a process to fix nitrogen from the air to provide artificial nitrogen fertilizers for the soil (receiving the Nobel Prize in Chemistry in 1918). This trailblazing technology made a tremendous contribution to increasing food production to support the increasing population worldwide. However, the same technology enabled the production of nitric acid from the atmosphere, which at that time was necessary for the production of explosives. This aspect created a connection between Haber and military technology. Using his comprehensive talents, Haber became absorbed in research on toxic gas as a weapon of massacre. Actually, chlorine gas, which is heavier than air, was first used intrench warfare in World War I (Ieper 1915), followed by the development of a more advanced series of toxic gases. It was said that Haber's first wife (Clara Immerwahr 1870–1915, Ph.D. in chemistry) committed suicide to protest Haber's beliefs and actions.

History also highlights the issue of ethics at the time of the development of the atomic bomb. The same logic used by Haber to justify the necessity of toxic gases was repeated, that is, the development of a strong technology helps to end war, there by decreasing the number of people killed in warfare. However, history further shows that this logic escalates the warfare engagement of both parties and sacrifices many innocent civilians. As slavery has been recognized widely as an absolute evil, even though that acknowledgment took considerable time, the recognition of warfare as an absolute evil is rising gradually.

Furthermore, scientific technology is stepping into an untapped field, the manipulation of life, posing the possibility of creating an android and requiring a new strong ethical perspective immediately. In biotechnology, the issue of ethics is increasingly important. In this paper, I will discuss the perspective of "Ethics-Based Engineering," including the future role of engineering academies worldwide.

2 Engineering and the Creation of Human Artifacts

Scientific technology, especially engineering, has a close connection to people's lives. Science refers to activities that unravel the world of nature, including humans, accurately and at a deeper level. On the other hand, engineering refers to creating human artifacts that do not exist in the world of nature. Both science and engineering need the support of technologies.

The origin of the word stem of *science,* “sci-,” is from the ancient Greek “skei-,” which means “split/divide.” Thus, the original meaning of science is considered to be “to understand nature by dividing or breaking down the system into elements,” which is linked to the Reductionism of René Descartes (1596–1650)—“to understand the essence of phenomena through analysis.” The word stem of *engineering* is “gin-(gen)” meaning “to give birth or create.” The suffix of *science,* “-en,” is attached after the word stem, whereas the “en-” of *engine* is the prefix attached to “gin-(gen),”meaning “to produce” a system to generate power to do an intended task using steam, gasoline or another power source. This concept applies not only to objects but also to information, as reflected in such a term as *search engine.* The term *engineering* always refers to activities to produce intended objects—human artifacts that do not exist in the world of nature.

The word stem of *technology* is “tekhne-,” which has the same root as the ancient Latin “ars.” In ancient Greek and Latin cultures, technology and art were considered to be similar activities. The three concepts of science, engineering and technology intersect orthogonally with each other according to their word stems. These are different from the mixed-up or multilayered general concepts of “scientific technology” and “art” being used in the 18th century and thereafter.

Because science is about uncovering natural phenomena and understanding the world of nature at a deeper level, the ethics of science is not highlighted and related observation is not required except for a few exceptions, whereas because engineering is about creating and producing human artifacts that are nonexistent in nature, the ethics of engineering inherently is an important issue.

3 Human Ethics

Remarks such as “God is with us” or “We follow God’s direction” are often heard from two countries or entities fighting each other. Both parties regard their God as the same, the one and only absolute existence. How can such logic be contradictory? Each party justifies its ideas and activities in the name of God, an absolute existence. Even though wonderful religions exist in the world, if their doctrines are used to justify what they want to do, ethics and morals do not exist in the real world. Unfortunately, human history shows the repeated use of such doctrines for wrong purposes.

Many innocent civilians have been killed by weapons created by the results of engineering. High-performance weapons for massacre are used by both parties in strife. The development and supply of such weapons is a serious issue.

Many Eastern concepts consider that humans are allowed to live in an eternal time and universe. If an absolute person or a great existence behind nature beyond human power is considered as God, such existence could be logically possible. Humans who are allowed to live a moment of the eternity on this planet need to live with compassion for others, complying with the rules of the world of nature. In such a way of life, we find various feelings, for example, joy and passion. To live in harmony with nature, ethics and morals are required. For engineering, which creates new human artifacts based on science and technology, ethics and morals are essential.

Unfortunately, in the world of science, the fabrication of scientific theses is seen because of intense competition. Because science involves human activities to uncover the world of nature accurately and at a deeper level, nothing is left if a fact is distorted. Ethics and morals are a compliance-based norm of human behavior and are nothing more than that as far as science goes. On the other hand, because engineering involves activities to create human artifacts that do not exist in the world of nature, the ethics and morals for engineering require a higher level of awareness and discipline than mere compliance with laws and regulations because the results of engineering could directly affect human lives. It is necessary to recognize clearly that more stringent ethics and morals are required from engineers than from scientists. The issues recognized by Nobel and even the Curies mainly concerned ethics in the domain of engineering.

4 Ethics of Engineering

Gene editing, which is currently a part of flagship research for regenerative medical treatment, is closer to engineering than science. How much should we be allowed to modify and/or manipulate humans? Traditionally, medical treatment was compared to a candle. A candle has a limited life. The candlelight eventually goes out when the candle wax is gone. In midlife, the wind could blow out the light and therefore must be controlled to sustain the light's life, which is analogous to the mission of medical treatment. This concept, which is the origin of medical treatments, is about to be changed by the greed of humans, who want to add unlimited amounts of candle wax. Humans eventually die. Because we all die, our lives shine, which has been observed by sages since ancient times.

The compliance required from society means observation of laws and regulations. Although compliance often overlaps with ethics, there are cases where they are not the same. Even without the violation of laws, there are various ethically unacceptable cases. Especially in the case of biotechnology, a rapidly developing science and technology, related laws and regulations are yet to be properly in place and related ethical issues have not been addressed fully. Increasingly important in the future will not be the compliance simply to win litigations but rather the ethical perspective based on the dignity of humans that helps distinguish what needs to be done and what does not need to be done. In particular, engineers who could potentially have direct impact on human lives and the global environment must not only comply with laws but also possess a refined sense of ethics.

Old Indian philosophy includes excellent insight, for example, "warmheartedness" related to the depth of the human mind that is not possessed by other animals. This concept, which had been passed down orally since before B.C. 500 and was later recorded in the ancient Pāli language, directly links to the dignity of humans and was adopted in Buddhism as *Shimuryoshin* ("unmeasurable broad mind"), consisting of the following four *Jihikisha* concepts: (1) *Metta* = "mercy," meaning mercy, kindness expecting no return, friendship or parental love-like affection; (2) *Karuna* = compassion, meaning feeling others' agony as one's own; (3) *Mudita* = "joy," meaning to enjoy others' happiness as one's own; and (4) *Upekkuha* = "equanimity" or composure not obsessed with self, objects and other persons (a comprehensive view from a higher-tiered stance). From the perspective of cognitive neuroscience, these concepts underlie the Theory of Mind, or the high-level ability to read others' minds.

Engineering academies worldwide need to address the significant issues of understanding the ethics for engineering, and the education and dissemination of such ethics.

5 Specific Ethical Issues of Engineering

In addition to addressing issues such as the development of weapons for massacre, global warming and biodiversity, there is an urgent need to address ethics related to biotechnology such as the rapidly advancing brain research.

A recent research paper revealed that a sense of guilt can be stopped by giving simple electric stimuli to the brain, that is, transcranial direct-current stimulation (tDCS; Haruno, M., *The Journal of Neuroscience*, 2015). Furthermore, memory can be strengthened or oblivion can be promoted. What a person is thinking can be measured externally, which is an area of ultimate privacy, posing infinite ethical issues in neuroscience (Koizumi, H., *SeimeiRinri (Bio-ethics)*, 2006).

Another recent issue posing a direct ethical question is gene editing. In the past, the engineering of embryonic stem cells became controversial in light of ethics. Even for the purpose of saving life, the question remains whether the use of embryonic stem cells with the potential to grow to an individual in the future should be permitted. Given such background, research using human embryonic stem cells has been managed carefully.

Therefore, expectations are high for induced pluripotent stem (iPS) cell technology, which has the potential to produce stem cells from somatic cells without using embryonic stem cells, for which Dr. Shinya Yamanaka (1962–) received the Nobel Prize in Physiology or Medicine in 2012. However, in an extreme case, stem cells could be created from a person's somatic cells or leukocytes using the iPS cell technology. In addition, gene editing helps create men and women with distinct individual characteristics. And it is no longer so farfetched that such created men and women could come of age, marry and produce a child. Considering such a future, the issue of ethics will continue infinitely. Stem cell technology, in particular, is typical engineering. The mechanism as to why normal somatic cells can be initialized by inserting four types of genes, which is a scientific aspect of iPS cell technology, has yet to be unraveled. Nevertheless, the Nobel Prize in Physiology or Medicine was awarded and the engineering to apply the mechanism is being promoted at a national scale, based on the concept that safe utilization of this technology will contribute to a healthy society and longevity, which is also economically beneficial, as well as human welfare and happiness. However, the question remains as to whether this is an ideal direction for future society to pursue. This is a typical example of the complex ethics of engineering.

The alchemy in the medieval era, as seen in the illustrations inserted in *Faust* by Johann W. von Goethe (1749–1832), not only converted base metal into precious metal but also had the ultimate purpose of producing an android ("homunculus" at that time). Furthermore, Isaac Newton (1642–1727), the forefather of modern physics, was secretly researching alchemy. That fact was revealed by the insightful John M. Keynes (1883–1946), an economist at the University of Cambridge, for which Newton was working, and later discovered in many documents written by Newton himself. Why was such a famous scientist as Newton secretly researching alchemy? A natural answer to this question could beth at Newton was well aware of the issue of ethics. At any time in history, the curiosity and pursuit of fame of scientists and engineers are endless indeed.

6 Conclusion: What Should Engineering Academies Worldwide Do About Ethics?

Engineering academies are professional groups of engineers with ample knowledge and experience and comprehensive perspectives. The academies' founding philosophies were neutral and independent of government and ideologies. Such organizations are extremely important, functioning as a compass to lead future engineering in a right direction, which cannot be done by other organizations. In fulfilling such a crucial role based on excellent intelligence, each country's engineering or scientific academy needs to prioritize addressing the issue of ethics.

As a recent trend of engineering and scientific academies, they are increasingly inclined to education. How should we educate the young people who will lead the future about ethics? To establish ethics education, what are the issues to address and what are the obstacles? The K-12 STEM (Kindergarten through 12th-grade Science, Technology and Engineering) education advocated making a future innovation ecosystem (platform) in the United States. I hope that would include how to nurture an ethical mind. Ethics education should be included not only at the engineering departments of universities or in lifelong education but also widely in general education. This matter needs to be discussed at each country's engineering and scientific academy, and the International Council of Academies of Engineering and Technological Sciences Inc. (CAETS) needs to focus on this point and formulate specific actions immediately.

Although engineering is extremely important, it is not the ultimate goal of humans. Engineering is a means for us to acquire *annei,* or human security (e.g., global warming, pandemics), on this planet and fulfill our quality of life. Our goal is to utilize engineering as a positive measure for humans and other life forms. Recently, in many fields the purposes and measures are getting mixed up. The mission of engineering and scientific academies, with their concentrated intelligence, is to be an ethical compass.

Every time a scandalous incident occurs, the same thing is repeated—redistributing materials indicating the ethical matters to be observed, addressing a measure that should have been taken, often without discussing the specific measures and the evaluation of such measures. From now on, however, ethical issues that involve the survival or extinction of the human race will increase, leaving no room for compromising ethical issues. We are entering an era that requires convincing ethics to ensure that the human artifacts to be created are conducive to the entire human race. Profound discussion as to the future direction is needed as part of the engineering academy's activities. I sincerely hope that the CAETS will pioneer a bright future for the engineering and technological sciences.

Acknowledgments I would like to express my sincere gratitude to Profs. Emma Rothschild and Amartya Sen of Harvard University and Dr. Mahathir bin Mohamad, who was the fourth Prime Minister of Malaysia, for their invaluable suggestions.

Scalable Medical Devices: Personal and Social

Ashutosh Sabharwal and Ashok Veeraraghavan

Abstract Health and well-being is poised to become more quantitative and connected, with a brand new class of devices. These personalized and population-scale health diagnostics, assessment and analysis devices will be markedly different from the current generation devices in three important attributes: (a) access, (b) affordability, and (c) required operational skill. We discuss the design principles and examples of next generation of medical devices and highlight the potential of this new class of devices to revolutionize health monitoring and eventual health outcomes.

Keywords Medical devices · Measuring behavior · Quantified communities

1 Introduction

Healthcare continues to be a significant and ever-growing challenge for all nations. The healthcare challenge is inherently multi-dimensional: variables such as access, affordability, quality, health behaviors and perception all contributing to the eventual health outcomes in complex ways. Effective healthcare solutions, as a result, require multi-disciplinary solutions. The engineering challenge is that of measurement, how do we measure bio-markers and (yet-to-be-discovered) "behavioral-markers," accurately. The question thus is *how do we then scale these measurement techniques so that they reach the target population, which in many ways is socially, economically and geographically diverse?*

While past decades have witnessed great advances in our ability to measure clinically relevant parameters, a large part of the design has assumed operation in a diagnostic lab with expert trained technicians. To increase access to healthcare, a

A. Sabharwal (✉) · A. Veeraraghavan
Scalable Health Initiative, Department of Electrical and Computer Engineering, Rice University, Houston, USA
e-mail: ashu@rice.edu
URL: http://sh.rice.edu

R. Shorey and P. Ghosh (eds.), *Healthcare Engineering*,
DOI 10.1007/978-981-10-3111-3_2

first crucial challenge is to reduce the cost and expertise barrier to conduct advanced diagnostic tests.

Ideally, a diagnostic test should be designed such that it can be performed by anyone and anywhere. However, some thought immediately reveals there are new challenges in developing diagnostic devices to meet the envisioned anywhere-anyone use. First, the operating conditions cannot be controlled, and hence the design has to fundamentally account for a much larger range of operating conditions. That is, the design has to be *operating-condition-robust*. Second, since the operator expertise is no longer guaranteed, the design should require *zero-training* such that the devices can be operated by almost anyone without any training. Third, the devices should seek to achieve *clinical grade accuracy*, to ensure that the data from the devices is diagnostically relevant. If a medical device achieves the three goals of operating-condition-robustness, zero-training and clinical-grade accuracy, it forms the foundation of a *scalable medical device*.

Adding to the diagnostic challenge, a significant challenge in healthcare is human behavior that impacts all aspects of our health. For example, it is well known that many chronic disease patients find it challenging to adhere to their medical regimens (see e.g. CDC data on chronic diseases like Asthma, COPD, diabetes). Analogously, many individuals find it challenging to adopt healthy lifestyles. Human behavior is complex and multi-dimensional, making it extremely challenging to understand the causal pathways that lead us to make specific choices.

While it is well appreciated in concept that our behavior is heavily influenced by our social context, it is still not possible to measure its strength and overall impact on the health for an individual. Product marketing and public health campaigns have long used social network ties to achieve their aims. However, we still do not have a systematic understanding of how to leverage our social networks to impact positive change for an individual.

Opportunity: The rise of smartphones has led to a multi-faceted opportunity to meet both challenges. First, there is a growing acceptance that more of our world will be smartphone-connected and app-controlled. Second, participation in online social networks has made us increasingly more comfortable with sharing personal data, at least in closed circles of our friends and families. Third, there is an increasingly more acceptance to behavior logging, where more of our actions can be automatically logged. The best example is perhaps activity monitors, like Fitbit and similar products, that automatically log and categorize physical activity and in some cases, related physiological state. Finally, the preponderance and ubiquity of smartphones, in all corners of the world, mean that smartphone-centric solutions may provide the potential to reach a geographically, socially and economically diverse population, resulting in at-scale monitoring and interventions. Our key hope is that the more than billion smartphones being sold world-over today, become the distributing health and wellness infrastructure that ushers in a new era of quantitative, connected and scalable health monitoring, diagnosis and interventions.

In the sequel, we divide our discussion into two classes of medical devices: (i) scaling traditional diagnostics, (ii) measuring social context.

2 Scaling Traditional Diagnostics

The key design goal is enabling measurements by anyone and anywhere. Thus, the devices have to be designed for scalability, to be used in non-ideal conditions and by non-expert operations and yet be clinically accurate. To appreciate the design thought process, we use two examples, CameraVitals and mobileVision, from our recent research that illustrate the concept of developing scalable medical devices. Other examples from our recent research include mobileSpiro for easier spirometry [1, 2], unconstrained gaze estimation for tablets [3], improved methods to manage Asthma [4] and even simplifying mHealth trials [5]. Our ongoing research also addresses zero-effort respiratory health status diagnostics, automated mental health tracking and measuring cognitive engagement.

Case Study 1—Rice CameraVitals: Our first design study targets measuring some of the most basic diagnostics—heart rate and breathing rate—in a manner that is much more scalable than current methods.

Current Practice All common techniques to measure the vital signs are based on contact sensors such as ECG probes, chest straps, pulse oximeters and blood pressure cuffs. Contact-based sensors are so prevalent that we do not stop to think about them. However, they are not convenient in all scenarios, e.g. contact sensors are known to cause skin damage in pre-mature babies during their treatment in a neonatal intensive care unit (NICU). Similarly, elderly may have difficulty time following precise instructions, thereby contributing to lower rates of regular self-monitoring. Finally, one of the challenges of contact-based vital sign measurement that they all require special hardware, thereby limited by the shipping, service and associated costs.

Scalable Design The driving question is how do we create a highly scalable method to measure vital signs. It is well known that software is simply more scalable than hardware, especially when considering all steps in the process, from creation to delivery. Currently, the most accessible computing platform is smartphones, and apps have made nearly everyone comfortable with using software. Ideally, if vital sign measurement was an app on the phone, using only its built-in sensor, then we have the foundation for a highly scalable approach.

Rice CameraVitals In [6], we presented a new method to turn video selfies using cameras into a vital sign estimator with clinical-grade accuracy. While using a camera to estimate vital signs has been known for a decade; see [6] for more details, all prior proposals have found it challenging to measure vital signs accurately for darker skin tones, in low ambient lights and in the presence of unavoidable natural motion. Our method, distancePPG, leveraged ideas from computer vision, motion tracking and wireless communications to develop a composite method that provides a significant gain in measurement accuracy, overcoming nearly all challenges.

Case Study 2—Rice MobileVision: Our second design study is that of retinal imaging, that falls more in the category of advanced diagnostics and hence is often performed in a dedicated lab or in specialist's office. Retinal imaging is a core diagnostic for diabetic retinopathy and macular degeneration. Beyond the core diagnostic value, retina is one of the few places in human body where we can directly image internal tissue, like blood vessels, with a *non-invasive* procedure, making it particularly suitable for regular screening.

Current Practice Current devices handle the above challenge with a combination of ideas. First, pupil response is removed from the picture by dilating the eyes (which in essence paralyzes the eye muscles). Second, to ensure that adequate light enters the eye, the patient and the imaging system have to be carefully aligned manually. Proper alignment ensures that sufficient light enters the eye. Third, an expert technician manages the whole process and takes the image at the right moment. In summary, all forms of motion (eye's reflex to light and patient's normal movement) and imaging moment are carefully and manually controlled.

Challenge The biggest challenge in imaging the eye is the eye itself. First, there are strict safe limits on the amount of light energy which can be projected into the human eye. Second, the pupillary response to visible light limits the amount of light that enters the eye. Lastly, only about 1 out of 1000 incident photons is reflected by the retina. Thus, not only we cannot project too much light into the eye, very little light comes out of it.

Rice MobileVision In [7], we presented a complete prototype design of a system that requires no user intervention in imaging, is voice activated and employs novel signal processing. The Rice mobileVision system consists of a goggles-like design that the patient wears over his face. The patient's personal smartphone attaches to the device, and acts as the sensor (camera), computing and communication platform. We have designed and developed two prototypes. The first prototype is intended for tracking the degradation in patients already diagnosed with retinal diseases and achieves a 50° field of view at 10 μm resolution but requires the acquisition of multiple images. The second prototype is designed to be a rugged, on-the-field, screening device for diabetic retinopathy and is a first-of-its-kind, self-administered retinal imaging device that achieves a 20° field of view and 20 μm spatial resolution.

3 Measuring Social Context

We start with a definition of social context or more equivalently social environment, put forth in [8]—"Human social environments encompass the immediate physical surroundings, social relationships, and cultural milieus within which defined groups of people function and interact. […] Social environments are dynamic and change over time as the result of both internal and external forces.[…]".

Social context can have a large impact on our health behaviors. For example, eating healthy requires access to healthy food choices and strategies to manage nutrition. Healthy food choices can vary significantly from one region to another, and nutritional strategies are often learnt from our social environment (homes, friends, schools). As we repeat our actions, the repetition of our specific behaviors gets coded as habits. That is, we may be inclined to eat a specific type of diet, primarily because it is part of habitual decision making. In short, habits are automatic behaviors, triggered by cues and are believed to form a bulk of our actions on day to day basis.

From public health perspective, it is highly desirable that the population adopts good habits (e.g. regular exercise, healthy foods) and avoids bad habits (e.g. smoking or sugary drinks). The challenge to achieving the desired public health outcome is well captured by the age-old adage "old habits die hard". Substantial evidence rooted in neuroscience [9] has shown that habitual behavior is supported by strong and stable mental representations, which makes forgetting or unlearning of habits very challenging.

Prior research has studied the role of incentives to change habits. One possible strategy is to offer financial incentives to either reduce or eliminate an undesirable behavior (e.g. smoking) or increase a desirable behavior (e.g. regular exercise). Many recent studies have shown promise with its approach. For example, the studies reported in [10] offered different level of financial incentives to university students to attend the gym. It was found that by requiring some students to attend the gym multiple times for a financial compensation led to increased gym attendance, even *beyond* the period of the study. However, while a later study [11] managed to replicate the results, it also discovered that there was a substantial decay in gym participation after the winter break.

A very promising result was reported in [12], where the researchers obtained a detailed friendship network from the participants, living in the same residence hall, before the start of the study and then provided financial incentives like the prior studies. A key finding was that the gym participation was not only increased by the financial incentive, but its effect was higher for the participants with friends who had also been incentivized. That is, the overall participation was higher for participants if their friends were also in the incentive group, and lower if their friends were in control group. The finding in [12] shows the importance of the social network in enhancing the effect of incentives for habit change.

While the result reported in [12] is highly intuitive and satisfying, the overall evidence to support the role, methods and strength of habit change due to social engagement remains in its infancy. *A key challenge lies is our ability to measure social context, and its impact on our specific health behaviors and outcomes.*

Rice OWLympics[1] At Rice, we are developing the engineering core to measure social context (e.g. our friendship network), health metrics and mental well-being using a combination of (i) sensors mounted throughout the campus that can measure the type of activity, its intensity and its impact on the person's vitals,

[1]OWLS is the name of Rice University's sports teams. The name, OWLympics, is thus derived from the Rice team name.

(ii) mental well-being using innovative analysis of one's digital footprint, and (iii) social network interactions, both in-person and online. The combined data will enable a unique *Quantified Organization,* complementing the quantified self movement, where we can leverage both the personal and social for our best health outcomes. The project is still in its early stages, and the results from the pilot program were reported in [13].

References

1. S. Gupta, P. Chang, N. Anyigbo and A. Sabharwal, *mobileSpiro: Accurate Mobile Spirometry for Self-management of Asthma,* Proceedings of First ACM Workshop on Mobile Systems, Applications, and Services for Healthcare (mHealthSys), 2011.
2. S. Nikkila, G. Patel, H. Sundaram, A. Kelliher and A. Sabharwal, *WindRunners: Designing a Game to Encourage Medical Adherence for Children with Asthma,* Proceedings of CHI, 2012.
3. Q. Huang, A. Veeraraghavan and A. Sabharwal, *TabletGaze: Unconstrained Appearance-based Gaze Estimation in Mobile Tablets,* submitted to IEEE Transactions on Image Processing, August 2015 (available at arXiv:1508.01244).
4. R. Biswas, P. Chang, H. Dharamasiri, G. Patel and A. Sabharwal, *AsthmaGuru: A Framework to Improve Adherence to Asthma Medication,* Proceedings of Wireless Health, 2013.
5. P. Washington, M. Kumar, A. Tibrewal and A. Sabharwal, *ScaleMed: A Methodology for Iterative mHealth Clinical Trials,* Proceedings of HealthCom, 2015.
6. M. Kumar, A. Veeraraghavan and A. Sabharwal, *distancePPG: Robust Non-contact Vital Signs Monitoring Using a Camera*, Biomedical Optics Express, Vol. 6, Issue 4, pp. 1407–1418, 2015.
7. A. Samaniego, V. Boominathan, A. Veeraraghavan and A. Sabharwal, *mobileVision: A Face-mounted, Voice-activated, Non-mydriatic "Lucky" Ophthalmoscope*, in *Proceedings of Wireless Health*, 2014.
8. E. Barnett and M. Casper, *A Definition of "Social Environment,"* American Journal of Public Health, 91(3), page 465, March 2001.
9. C. Roberto and I. Kawachi, *Behavioral Economics and Public Health,* 2015.
10. G. Charness and U. Gneezy, *Incentives to Exercise*, Econometrica, 77(3), pp. 909–931, 2009.
11. D. Achland and M. Levy, *Habit Formation and Naivete in Gym Attendance: Evidence from Field Experiment,* Unpublished work.
12. P. Babcock and J. Hartman, *Exercising in Herds: Treatment Size and Status Specific Peer Effects in a Randomized Exercise Intervention.* Unpublished work.
13. R. Biswas and A. Sabharwal, *OWLympics*, Poster at SAMSI, 2013.

Therapeutic Platforms for Ischemic and Traumatic Brain Injuries Across National-Level Neuroimaging Grids

Prasun K. Roy, Suhela Kapoor, Vikas Pareek and V.P. Subramanyam Rallabandi

Abstract We delineate feasibility of computational clinical trial platform for therapy optimization approach to neuroprotection and neurorestoration in cerebrovascular accident or intracranial injury. Cell kinetics or drug response characteristics is quantified by systems biology model that is validated based on experimental findings of animal and human study. Platform is built on the Government's high-speed NKN system, connecting medical centres across the country.

Keywords Regenerative medicine · Stroke · Traumatic brain injury · Treatment planning · Multicentric trial · Grid computing

1 Introduction

Brain injury can occur due to brain damage produced by various means as by cerebrovascular accidents (like insults from haemorrhage, aneurysm, thrombosis), or by mechanical accidents (traumatic brain injury, violence or assaults, sports, and vertical falls). In 5 years time, brain injury becomes the leading disease burden globally, after myocardial infarction and malignant disease, with India's burden as highest [1–3].

While customary symptomatic clinical management are available, there is much need of newer therapeutic approaches, since currently no adjunct treatment approaches are available for appreciably improving neurorestoration or neuroregeneration,

---

S. Kapoor, V. Pareek, V.P. Subramanyam Rallabandi—Equal contribution.

P.K. Roy (✉) · S. Kapoor · V. Pareek · V.P. Subramanyam Rallabandi
National Brain Research Centre (Deemed University),
Ministry of Science and Technology, Government of India,
Manesar 122051, NCR Delhi, India
e-mail: pkroy@nbrc.ac.in

R. Shorey and P. Ghosh (eds.), *Healthcare Engineering*,
DOI 10.1007/978-981-10-3111-3_3

except the scope of enhancing the internal or endogenous neural stem cell stores and niches in the adult human brain [4].

There looms a chronic disease pandemic both in developing and developed nations, and India has the world's highest disease burden (700 Million DALY-units), costing US$ 17.4 Trillion during 2012–2030 [3]. A major global challenge is drawing meaningful findings from the enormous clinical, genomic and imaging data of the investigations on stroke and traumatic injury, from multicentric clinical trial perspective [5].

These problems can be dealt with by taking recourse to multicentric brain imaging grids that enable scientific, medical and engineering investigators to collect, archive and analyze large amounts of imaging data. These grids also enable the investigators to access suitable resources for enabling computationally intensive data analyses, and thereby develop improved approaches and models of disease understanding and treatment planning.

2 Methods

We explore the possibility of developing a neuroinformatics and neuroimaging-aided computational methodology for delineating endogenous neural stem cell generation and neuroprotection, under activation by growth factors or drugs with similar effect. This enables a restorative and regenerative approach for recuperating the damage under traumatic injury and stroke.

We have elucidated a systems biology approach to the neuroregenerative therapeutic approach to stroke and traumatic injury [6, 7] which we further expand and extend here. Furthermore, we also probe the systems integration of the aforesaid procedure to facilitate multicentric international cooperation utilizing the Brain Imaging Grid approach, based on standardization protocols [8] developed by Montreal Neurological Institute, McGill University.

Utilizing the Brain Imaging Grid network, we elucidate the optimization of neuroprotective therapy of Brain Injury (BI), for this we adapt the large data analysis and infrastructure methodology delineated by the International Neuroinformatics Coordinating Facility, namely the INCF initiative of Centre-BI [9]. For synthesizing the above methodology, we develop a novel integration of Multiscale neuroinformatics approach, applying it to the optimization of neurorestorative and regenerative efficiency as the dose and time variables of the growth factors (concentration of drugs) are varied.

It may be mentioned that we have earlier pursued cooperative neuroimaging studies involving on investigating the neurobiology and biomarkers of epilepsy involving collaborators from multiple national centres [10] and of human ageing-induced senescence, degenerative and cognitive processes incorporating collaborators from international centres [11]. In this report, we explore the suitability of a multicentric platform approach to investigate brain damage, in stroke and traumatic injury.

For marshalling the methodology, we utilize the procedure of Multiscale approach to systems physiology and systems pharmacology [12], which is the emergent new approach for the handling and integrating the complexity in brain studies, which interlinks information across all scales of the brain, from biochemical molecules, genes, cells, tissues, to functional networks, behaviour, drug action, and psychometry, so as to develop a complete picture of the brain system.

3 Results

We here particularly consider cerebral hemispheric injuries of stroke and head injury. Using quantitative cell-growth analysis, we develop a computational formulation of intracranial cell proliferation (gliogenesis, dendritogenesis, angiogenesis), based on experimental findings, in both preclinical and clinical setting, i.e. the adult rodent system and adult human system, post-injury.

We analyse the histologically-demonstrated frontal migratory channel-system of stem cell generated endogenously in the ventricular zone and proceeding toward the stroke or traumatic injury sites in the cerebral hemispheres, via two migratory channels which are:

(i) In rodent: Lateral cortical migratory channels, or
(ii) In human: Medial cortical migratory channel.

Using a digital spatiotemporal simulation platform, we delineate a quantitative cellular kinetics and migration model to gauge the regeneration intensity of stem cell or neuroblast, which alters with the age of the organism (animal or human). The model then furnishes the generation of mature glial cells/neuronal cells, vis-a-vis the dose-time concentrations of the drugs or growth factors, thus a computational clinical trial platform is developed.

To enable implementation across multiple centres across a country, we now consider the feasibility of the setting up of the above-mentioned digital spatiotemporal simulation platform, integrated with the national-level Brain Imaging Grids and Neuroinformatics databases/networks across different regions, that have been earlier commissioned by national and international agencies, such grids being as:

(i) Europe (Neugrid, NeuroLog),
(ii) Asia (APGrid, Pacificnet),
(iii) North America (C-Brain, BIRN),
(iv) Oceania (MedGrid, AIBL).

For adapting to the Indian context, we then endeavour to integrate the optimal therapy design platform of stoke/head injury with the Multicentric Brain Imaging grid for India with nodes in the 4 zonal regions, as;

(a) North (Delhi NCR): National Brain Research Centre; All-India Instt. of Medical Science.
(b) South (Bangalore): National Institute of Mental Health & Neurosciences.
(c) East (Calcutta): Institute of Postgraduate Medical Education & Research.
(d) West (Bombay): King Edward Medical College and Research Institute.

This platform uses the National Knowledge Network (NKN) infrastructure, set up the Ministry of Information Technology, having a connectivity speed of 100 Mbps. As a specific case for Brain Grid, we develop a neuroinformatics architecture in Linux-Apache-MySQL-PHP bundle, using the Government's National Knowledge Network (gigabit digital-grid of 1200 institutes/universities/hospitals) and CBrain prototype, with global linkage via Trans-Eurasia Network/TIEN-3.

4 Discussion

We obtain the definitive formulation of a quantitative model of the progenitor cell formation rate across the migration channels, undergoing proliferation kinetics. We then validate the formulation using available findings from:

(i) MRI and Neurological investigation: In recovery in human cerebral thrombosis and traumatic damage,
(ii) Immuno-histochemical study: In rodent experimental stroke and cerebral injury.

The enhanced recovery was tracked under defined dose of regeneration-promoting drugs as the neuropoeitins, e.g. carbamylated-erthropoeitin or glypromate. Our computational neurobiology formulation functions as an in silico clinical trial platform and correctly predicts and tracks the recovery endpoints, in both rodent and human systems, within 10 % error.

A bioinformatics platform is formulated for calculating optimum dose-time protocols of the neurorestorative drugs as neuropoeitins or fluoxetine, so that there is most effective clinical recovery performance, as estimated by Rankine score or sensorimotor grasping force. This enables a proper translational applicability for planning therapy designing in traumatic injury or cerebral stroke.

The promising potentiality of bioengineering and cell kinetics strategies to intervene at the early stages of brain ageing disorders, offers a considerable hope for the future in the ensuing demographic ageing pandemic. The infrastructure thus designed is expandable to other medical applications and harnesses a strong collaboration between neuroinformaticians, scientists, physicians and engineers.

We utilize the affordable country-wide India Brain Imaging Grid (using multi-scale modelling operating on the e-health sector of the National Knowledge Network, Ministry of Information Technology, Government of India), whereby we enable rapid multi-centric collaboration and optimization of disease-modifying intervention, with precision applications to therapy planning in stroke and intracranial injury.

5 Conclusion

We show the feasibility of developing a neuroinformatics platform for translational applicability for formulating proper optimal therapy of cerebrovascular accident and traumatic intracranial injury. The cellular kinetics and pharmacological response characterization can be quantified using a neurocomputational model, which is validates based on experimental and clinical finding of both animal studies and human investigations. Using available brain grid systems that connects various clinical neuroscience centres in different regions of a country, a standardized protocol for optimal therapy planning can be performed.

This endeavour utilizes online using remote connectivity with a centralized neuroinformatics platform for standardized image acquisition and treatment monitoring. We explore further integration with the development of:

(i) the multiscale neuroimaging ontology (International Neuroinformatics Coordination Initiative) and
(ii) the digital Brain Grids in various regions (E.U.Neugrid, MNIgrid and Asian Medgrid systems).

The prospects and enhancement of the methodology are dealt within the Indian scenario.

Acknowledgments The work of PKR, SK, VP and VPSR were respectively supported by Govt. of India's Dept. of Biotechnology, DBT (Glue Grant scheme and HRD scheme), Dept. of Electronics & Information Technology (Spatiotemporal mapping project), and Defense Research & Development Organization.

The cooperation of Alan Evans and Samir Das, McGill University, is greatly appreciated. Support for the logistics of the work is thankfully solicited from Tata Innovation Program, DBT, and Office of Principal Science Adviser, National Knowledge Network, Government of India.

References

1. A. DiCarlo, "Human and economic burden of stroke," Age & Ageing, 38, 4–5 (2009).
2. Lancet Editorial. "The changing landscape of traumatic brain injury research," Lancet Neurology, 11, 651 (2012).
3. Institute of Health Metrics and Evaluation. "Annual Report on Global Burden of Disease (GBD)", University of Washington, Seattle (2013).
4. C. Borlongan & D. Hess, "New hope for stroke patients: Mobilization of endogenous stem cells", Canadian Medical Association Journal, 174(7), 954–955 (2006).

5. J. Bjaalie. “Understanding the Brain through Neuroinformatics”, Frontiers in Neuroscience 2 (1), 19–21 (2008).
6. S. Kapoor, VPS Rallabandi, PK Roy. “Optimizing Performance of Neural Stem Cell Therapy of Ischaemic Stroke”. Frontiers of Neuroinformatics 7 (2014). doi:10.3389/conf.fninf.2014.08.000126.
7. V. Pareek, VPS Rallabandi, S. Kapoor, PK Roy. “Harnessing Neuroprotection and Regeneration in Traumatic Brain Injury: A Translational Computed Neurobiology Platform with Preclinical-Clinical Substantiation”, Foundation Day Conference, Translational Health Science & Technology Institute, New Delhi (2014).
8. S. Das, A. Zijdenbos, J. Harlap, D. Vins, A. Evans. “LORIS: A web-based data-management system for multi-center studies”, Frontiers of Neuroinformatics, 5 (2012). doi:10.3389/fninf.2011.00037.
9. A. Maas, D. Menon, E. Steyerberg, G. Citerio, F. Lecky, G. Manley, S. Hill, V. Legrand, A. Sorgner. “Collaborative European NeuroTrauma Effectiveness Research in Traumatic Brain Injury”, Neurosurgery, 76(1), 67–80, 2015.
10. S. Datta, S. Chakraborty, S, Mulpuru, B. Tiwary, N. Chakrabarti, PK Roy. “MRI characteristics of temporal lobe epilepsy using rapidly-measurable spatial indices with hemispheric asymmetry and gender features”. Neuroradiology, 57(9), 873–86 (2015).
11. W. Ottey, E. Diesen, R Ramaswamy, Subhadip Pal, VPS Rallabandi, PK Roy. “Ageing Alterations in Whole-Brain Networks during Adulthood Mapped with Minimum Spanning Tree Indices: The Interplay of Density, Connectivity Cost and Life Time Trajectory”, NeuroImage: Journal of Brain Function, 109, 171–189, 2015.
12. P. Hunter & T. Borg. “Integration from proteins to organs: The Physiome Project”, Nature Reviews Molecular Cell Biology, 4, 237–243 (2003).

Next Generation Devices and Technologies Through Regenerative Engineering

Cato T. Laurencin and Lakshmi S. Nair

Abstract The next decade will see pioneering research to regenerate musculoskeletal tissues. This will enable us to move from an era of advanced prosthetics, to what is termed as "Regenerative Engineering" which aims to regenerate complex tissues and organ systems. The paper presents an overview of the regenerative toolbox under development to address grand challenges in musculoskeletal regeneration.

Keywords Autografts · Pore-forming polyphosphazenes · Ligaments · Biocompatibility · Tissue transplants

C.T. Laurencin (✉) · L.S. Nair
Department of Orthopedic Surgery, Institute for Regenerative Engineering, the Raymond and Beverly Sackler Center for Biomedical, Biological, Physical and Engineering Sciences, University of Connecticut Health, Farmington, CT 06030, USA
e-mail: Laurencin@uchc.edu

C.T. Laurencin
Department of Chemical and Biomolecular Engineering, Biomedical Engineering, Institute of Material Science, University of Connecticut, Storrs, CT 06269, USA

C.T. Laurencin · L.S. Nair
Department of Material Science and Engineering, Biomedical Engineering, Institute of Material Science, University of Connecticut, Storrs, CT 06269, USA

C.T. Laurencin
Department of Craniofacial Sciences, School of Dental Medicine, University of Connecticut Health, Farmington, CT, USA

C.T. Laurencin
The Raymond and Beverly Sackler Center for Biomedical, Biological, Physical and Engineering Sciences, University of Connecticut Health, Farmington, CT, USA

R. Shorey and P. Ghosh (eds.), *Healthcare Engineering*,
DOI 10.1007/978-981-10-3111-3_4

1 Introduction

Musculoskeletal tissue damage occurs due to trauma, birth defects, aging and disease. In the United States alone, over 34 million musculoskeletal tissue repair surgeries are performed annually [1]. Current treatment strategies include physical therapy, use of biologically active agents or drugs and surgical intervention. Surgical intervention utilizes prosthetic devices or tissue transplants to regain the function of the damaged tissue. The past few decades have seen significant developments in orthopaedic prosthetic devices, however, with serious short-comings such as lack of sensory feedback and biocompatibility issues [2]. Current tissue transplants involve autografts and allografts. Among these, autograft is considered to be the gold standard due to its excellent healing ability and biocompatibility. The limited availability of autografts and less than optimal performance of allografts led to the emergence of next level of clinical interventions. Instead of repairing or replacing the damaged tissue and organs, the current efforts are focused towards regenerating them [3]. In this respect, the past 25 years have seen significant advances in tissue engineering approaches. We have defined tissue engineering as the "application of biological, chemical and engineering principles toward the repair, restoration and regeneration of tissues using biomaterials, cells and factors alone or in combination" [4]. By now the large body of research using these combination strategies have demonstrated the possibility to regenerate individual tissues such as bone and skin. However, the greatest medical challenges lies in regenerating complex tissues and organ systems. From a musculoskeletal perspective, this translates to regenerating a functional knee or whole limb routinely caused by modern life trauma, chronic diseases and aging. Addressing these grand challenges necessitates a paradigm shift, and requires a trans-disciplinary approach. This has led to the emergence of a field which we termed as "Regenerative Engineering" and has been defined as the "Convergence of Advanced Material Sciences, Stem Cell Sciences, Physics, Developmental Biology and Clinical Translation for the regeneration of complex tissues and organ systems" [5]. Studies so far demonstrated significant advances in these individual areas and the convergence approach will enable us to develop a regenerative tool box to achieve substantial and accelerated strides in regenerating complex tissues and organ systems. For instance, we now have a better understanding of adult and embryonic stem cells and more importantly the possibility of developing patient derived induced pluripotent stem cells with tremendous regenerative ability. From a biomaterials perspective, even though tissue engineering has used traditional degradable polymers and ceramics to develop three-dimensional scaffolds to repair tissues, the research in the past decade led to significant advances in biomaterial science in terms of custom designed biomimetic biomaterials [6]. Studies so far have also established the potential of these biomaterials to serve as unique delivery systems for effector molecules and growth factors in a spatial and temporal controlled manner to promote the regenerative process. These developments in biomaterial science are further supported by the tremendous developments in fabrication methodologies that harness the unlimited potential of nano and micro technologies. The use of nano and micro-structured scaffolds has increased our understanding of cellular behavior on extracellular mimic matrices and gave substantial insights to modulate cellular functions using advanced biomaterials. These combined with the emerging additive manufacturing techniques are opening the door to develop patient specific tissue or organ structures to support the complex regenerative processes.

Combining multipotent stems cells with inducible biomatrices holds the promise to regenerate complex tissues and functional organ systems. Regenerative engineering uses bottom up and top down approaches to address this grand challenge [7]. The bottom up approach uses a cell based approach by utilizing information derived from regenerative and developmental biology. Many lower amphibians still retains a remarkable regenerative capability mediated by the proliferating and differentiating cells in blastema. Studies have shown that the biological processes in the blastema is controlled by the spatial and temporal expression of various morphogenetic cues. Even though the humans have only limited regenerative ability compared to these lower amphibians, the knowledge we get from the regenerative process in the amphibians will inform the design of novel strategies to increase the regenerative potential of adult human beings [8]. This when combined with top down approaches could result in successful regenerative engineering methods to develop next generation devices and technologies. The top-down approaches on the other hand attempts to integrate cell/molecular biology, physics, materials science, and engineering. The successful approaches may involve universal platforms that are composed of biomimetic matrices and or scaffold systems, biologically active proteins and peptides as well as inducerons (small molecule inducers of cell differentiation) and stems cells with high regenerative potential.

The research conducted in our laboratory and that of others have led to the development of tools to engineer single tissues. Some of the successful strategies we have developed by now to regenerate musculoskeletal tissues such as bone and ligament are discussed below. Even though biodegradable polyesters have been extensively investigated, we designed and synthesized novel polymeric materials to provide versatile alternatives to these more readily available materials that have drawbacks due to their inherent design. One example is a family of biocompatible and biodegradable polyphosphazenes developed as unique polymer class with tunable properties to develop scaffolds to regenerate various musculoskeletal tissues [9]. Figure 1 shows the feasibility to develop polyphosphazenes that can undergo a unique polymer erosion process allowing the feasibility to convert from a solid matrix to a porous three-dimensional structure in situ and thereby can support robust tissue ingrowth [10].

Critical parameters impact the design of novel three dimensional scaffolds which in turn affect their functional outcome. For instance, we have developed mechanically competent three dimensional structures via sintering of polymeric microspheres as scaffolds to support load bearing bone regeneration. One key parameter that needs to be addressed while developing large scale scaffolds is the ability to vascularize the construct to support biochemical transport into their interior. Studies have shown that blood vessel invasion will occur only to a depth of several hundred micrometers from the surface of the scaffold. We have evaluated the efficacy of sintered microsphere

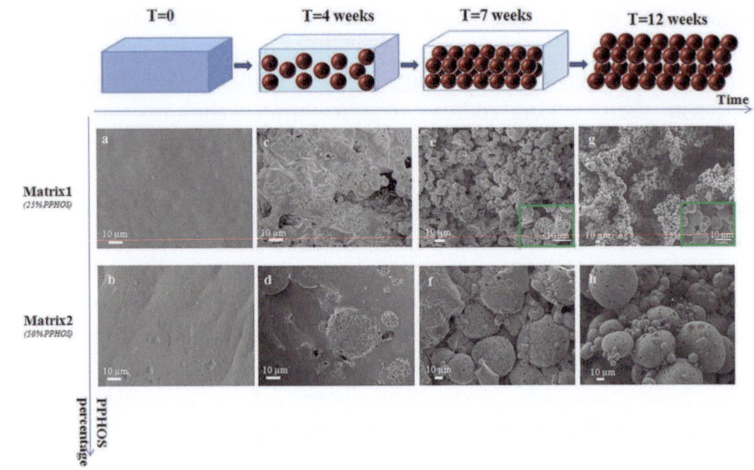

Fig. 1 SEM showing the morphologies of in situ pore forming polyphosphazenes as a function of time. Adapted from Ref. [10]

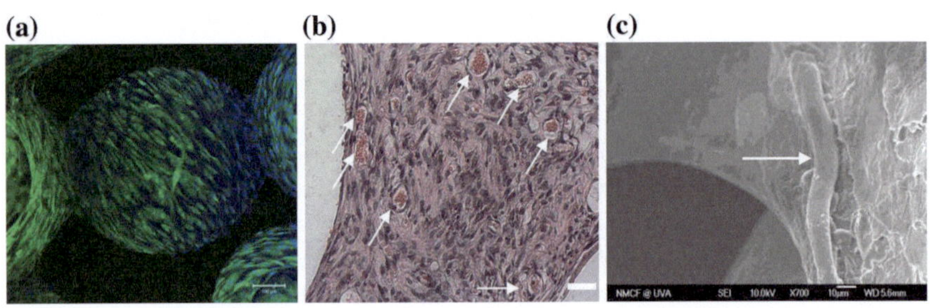

Fig. 2 **a** Expression of VEGF by cells seeded on sintered microsphere scaffolds. **b** Histology showing blood vessels identified by the presence of luminal structure containing red blood cells. **c** SEM micrograph showing blood vessel growth into the matrix. Adapted from Ref. [11]

(SM) scaffold to support endothelial cell proliferation and vessel formation (Fig. 2). The study demonstrated the feasibility of increasing vascular invasion in three-dimensional large constructs using a cell-based approach [11].

The versatile sintering process also allows for the development of porous structures of any shape and size. Figure 3a shows the feasibility to develop tubular constructs with the longitudinal axis void and a median pore size that closely mimicked the architecture of native bone by the microsphere sintering process. As can be seen from Fig. 3b, c, the tubular poly(lactide-co-glycolide) (PLGA) scaffold supported robust formation of osteoid (stained blue in 3B) and mineralized tissue (stained red in 3C). The dotted yellow line in the figure marks the original tube area showing complete bone formation [12]. The SM matrices are currently marketed as "MicroFuse®" (bone void filler with optimized strength and porosity) by Globus

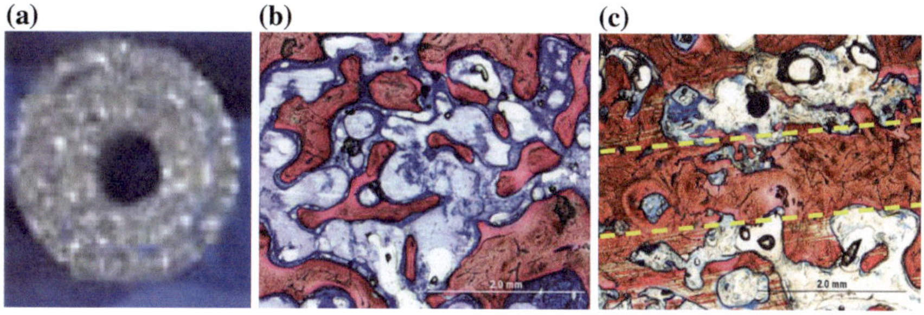

Fig. 3 **a** PLGA tubular sintered microsphere matrices. **b**, **c** Histological images at 24 weeks post implantation. Adapted from Ref. [12]

Medicals as a part of their minimally invasive, less disruptive and earlier recovery spine products [13].

The anterior cruciate ligament (ACL) serves as the primary stabilizing structure of the knee and the rupture of the ACL can cause both short and long-term disability. Due to the limitations of biological grafts, synthetic non-degradable ligament replacements have been explored. These prosthetics do not regenerate and hence lead to long term implant failure. ACL presents very unique challenges in terms of regeneration due to its complex hierarchical structure, its mechanical and structural properties, its bone-tissue interface, its intra-articular region and intra-synovial environment. We have developed a regenerative engineering approach towards ACL regeneration which offers the ability to provide an off the shelf implant for clinicians with the potential for initial mechanical support to allow early patient recovery. To provide initial biomechanical support and assist complete ACL regeneration, a biomimetic and biodegradable three dimensional braided fibrous scaffold was developed that have sufficient mechanical strength and an open porous, interconnected structure to promote cellular infiltration and formation of ligamentous tissue [14]. Figure 4a shows the three dimensional braided scaffold developed from poly(L-lactide) (PLLA). The hierarchical structure was created using braids with three regions: a femoral tunnel attachment site, an intra-articular zone and a tibial tunnel attachment site (Fig. 4b–d). To encourage ligament and bone in-growth and promote vascularization in these different regions, the fiber

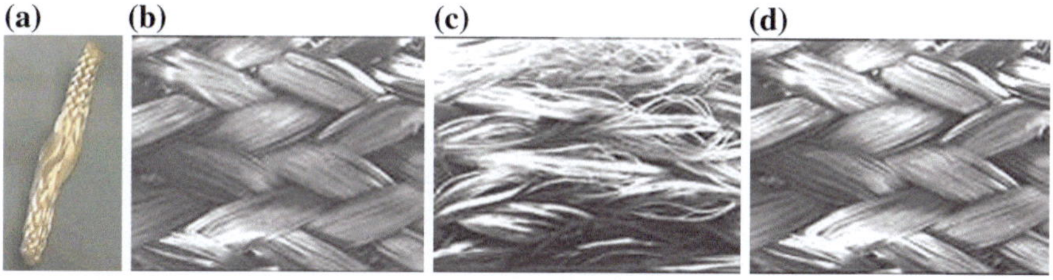

Fig. 4 **a** The fibrous hierarchy of the braided fibrous ligament scaffold. **b–d** SEM showing the three regions of the braided ligament scaffold. Adapted from Ref. [14]

orientation was varied to cause changes in pore sizes. In addition, the braiding process creates a structure with a continuous interconnected pore structure and increased available surface area for cell attachment that leads to enhanced regenerative response by allowing tissue ingrowth throughout the matrix. The in vivo study using a rabbit ACL model demonstrated the potential of the excellent healing and regeneration of ACL using the braided scaffold [15]. This was followed by a large animal study using an Ovine model [16, 17]. The study demonstrated the complete regeneration intra-articularly of the ACL using the degradable synthetic polymeric scaffold with hierarchical structure (Fig. 5a–d). This was achieved with good bony integration at the femoral and tibial regions (Fig. 5e–f). The

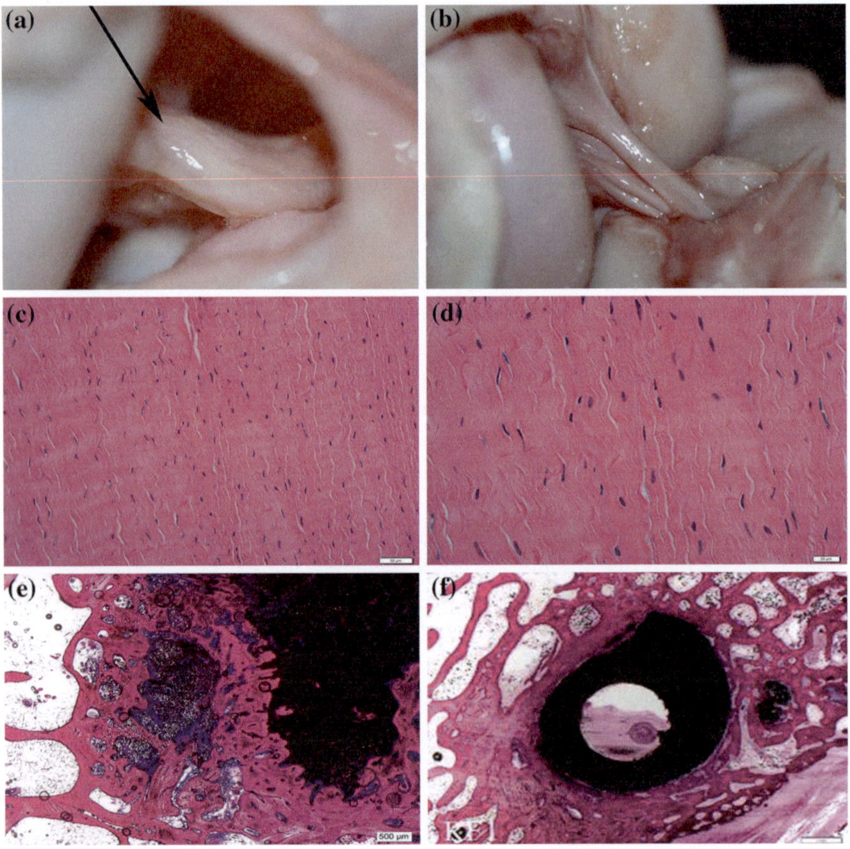

Fig. 5 Gross observation of the regenerated ligament **a** autograft **b** bioengineered ligament 3 month post implantation; **c** intra-articular region of the bioengineered ligament after ~1.5 years post implantation **d** high magnification image showing the crimp pattern and presence of fibroblasts; **e** histology of the proximal portion of the bioengineered ligament tibial tunnel at 1 year post implantation showing bone integration; **f** histology of the proximal portion of the bioengineered ligament femoral tunnel in the outside/in fixation at 6 month post implantation showing good bony integration. Adapted from Ref. [17]

regeneration of healthy tissue and complete resorption of the bioengineered graft by one year and a half indicated that the braided bioengineered ligament can function as a biodegradable scaffold to support ligamentization. The feasibility to achieve biomechanical properties similar to autografts during the remodeling phase demonstrated the ability of the bioengineered graft to function as a biomimetic scaffold [16]. The bioengineered braided PLLA scaffold is currently undergoing clinical trials in Europe to evaluate its efficacy in humans.

2 Conclusions

The research in the past few decades have demonstrated the feasibility to regenerate individual musculoskeletal tissues such as bone, ligament and cartilage. The unprecedented developments in stem cell biology, nanotechnology, biomaterial science and bioactive molecules have greatly contributed to develop these engineered tissues. Moving forward, we believe that the convergence approach puts forward by the field of Regenerative Engineering will facilitate a holistic and comprehensive approach to develop novel translational strategies to support complex tissue regeneration.

Acknowledgments The work is supported by the NIH Directors Pioneer Award (DP1OD019349). Dr. Laurencin was also the recipient of Presidential Faculty Fellow Award from the National Science Foundation.

References

1. Deng M, James R, Laurencin CT, Kumbar SG. Nanostructured polymeric scaffolds for orthopaedic regenerative engineering. *IEEE Trans. Nanobiosci.* 11(1), 3–14 (2012).
2. Talmo CT, Aghazadeh M, Bono JV. Perioperative complications following total joint replacement. *Clin. Geriatr. Med.* 28(3), 471–487 (2012).
3. Khademhosseini A, Vacanti JP, Langer R. Progress in tissue engineering. *Sci. Am.* 300(5), 64–71 (2009).
4. Laurencin CT, Ambrosio AA, Borden MD, Cooper JA. Annual Review of Biomedical Engineering. Yarmush ML Ed (Annual Reviews, Palo Alto); 1999; pp. 19–46.
5. Laurencin CT, Khan Y. Regenerative Engineering, Sci Transl. Med 4, 160ed9 (2012).
6. Guerette PA, Hoon S, Seow Y, Raida M, Masic A, Wong FT, Ho VHB, Kong KW, Demirel MC, Francesch AP, Amini S, Tay GZ, Ding D, Miserez A, Accelerating the design of biomimetic materials by integrating RNA-seq with proteomics and materials science. Nature Biotechnology, 31: 908 (2013).
7. Reichert WM, Ratner BD, Anderson J, Coury A, Hoffman AS, Laurencin CT, Tirrell D. 2010 panel on the Biomaterials grand challenges. J Biomed Mater Res A. 96: 275 (2011).
8. Gardiner DM, Bryant S, Muneoka K. Engineering Limb regeneration: Lessons for Animals that can Regenerate. In Regenerative Engineering (Laurencin CT and Khan Y Eds). CRC Press, Talor & Francis (2013).

9. Nair LS, Khan Y, Laurencin CT. Polyphosphazenes. In Introduction to Biomaterials (Hollinger Ed). CRC press, Taylor & Francis (2012).
10. Deng M, Nair LS, Nukavarapu S, Kumbar SG, Jiang T, Weikel AL, Krogman NR, Allcock HR, Laurencin CT. In Situ porous structures: A unique polymer erosion mechanism biodegradable in biodegradable dipeptide-based polyphosphazene and polyester blends producing matrices for regenerative engineering. Adv. Funct. Mater 20: 2794 (2010).
11. Jabbarzadeh E, Starnes T, Khan Y, Jiang T, Wirtel AJ, Deng M, Lv Q, Nair LS, Doty SB, Laurencin CT. Induction of angiogenesis in tissue-engineered scaffolds designed for bone repair: A combined gene therapy and cell transplantation approach. PNAS, 105: 11099 (2008).
12. Kofron MD, Griswold A, Kumbar SG, Martin K, Wen X, Laurencin CT. The implications of polymer selection in regenerative medicine: A comparison of amorphous and semi-crystalline polymer for tissue regeneration. Adv Funct Mater 19: 1351 (2009).
13. http://www.globusmedical.com/microfuse-technology-engineered-for-optimal-strength-and-porosity/.
14. Cooper JA, Lu HH, Ko FK, Freeman JW, Laurencin CT. Fiber based tissue engineered scaffold for ligament replacement: design considerations and in vitro evaluation. Biomaterials 26: 1523 (2005).
15. Cooper JA, Sahota JS, Gorum WJ, Carter J, Doty SB, Laurencin CT. Biomimetic Tissue-engineered anterior cruciate ligament replacement. PNAS. 104: 3049 (2007).
16. Laurencin CT, Walsh W, Bertolla N, Poggie R, Reilly J, Nair LS. An engineering solution to anterior cruciate ligament regeneration using a biodegradable and biomimetic matrix. ABJS, proceedings of the 2013 Association of Bone and Joint Surgeons meeting, Istanbul, Turkey, April (2013).
17. Mengsteab PY, Nair LS, Laurencin CT. The past, present and future of ligament regenerative engineering. *Regen. Med.* doi:10.2217/rme-2016-0125 (2016) (Epub ahead of Print).

Rising Healthcare Spending: Is Technology the Solution?

Pierre-Etienne BOST

Abstract In most developed or developing countries, healthcare costs, although largely variable when reported per capita (Fig. 1) are increasing continuously. Technology is often considered as the main culprit of these spendings: it is well acknowledged that new therapies are more expensive, new diagnostic tests, new surgical implants or devices are often costly.

Keywords DNDi · Drug for Neglected Disease initiative · DOMOHEALTH · NGS · HIV mangement · Healthcare cost

1 Introduction

In most developed or developing countries, healthcare costs, although largely variable when reported per capita (Fig. 1) are increasing continuously. Technology is often considered as the main culprit of these spendings: it is well acknowledged that new therapies are more expensive, new diagnostic tests, new surgical implants or devices are often costly. This is the main difference with Information and Communication Technologies (ICT). «Computers make things better and cheaper. In health care, new technology makes things better, but more expensive» says Jonathan Gruber, an economist at MIT who leads a health-care group at the National Bureau of Economic Research [1].

With aging population and urban concentration, chronic diseases but also emerging diseases are significantly growing. Some long-term degenerative but also acute diseases are more and more unendurable for various healthcare systems or health insurance policies. But the sooner these pathologies are diagnosed, treated or even prevented, the better it results in health economy and represents real opportunities for healthcare companies or new technologies.

P.-E. BOST (✉)
GENETHON, DNDi Scientific Advisory Committee Chair,
National Academy of Technology of France (NATF), Melbourne, Australia
e-mail: pierrebost@wanadoo.fr

R. Shorey and P. Ghosh (eds.), *Healthcare Engineering*,
DOI 10.1007/978-981-10-3111-3_5

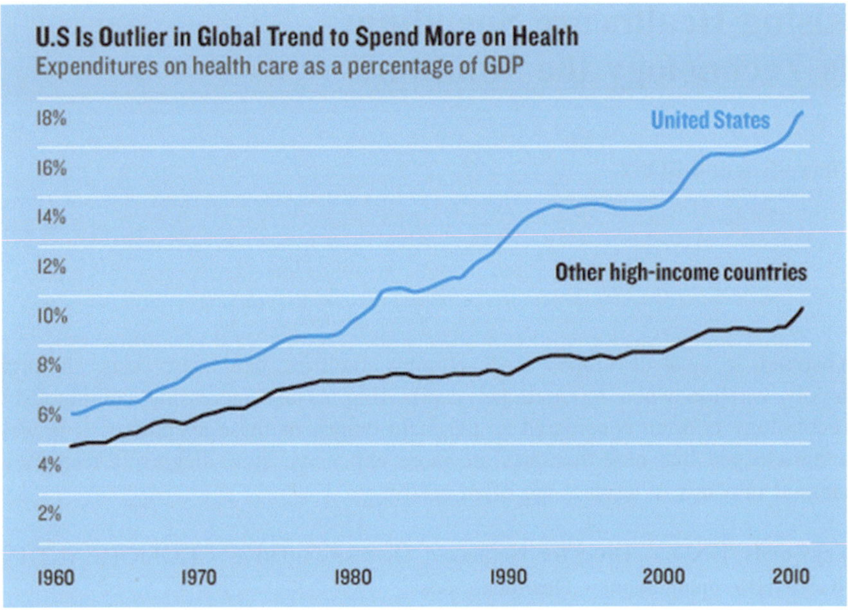

Fig. 1 Health care expenditure. *Source* DECD Health Data 2011

2 A New Paradigm for Healthcare R&D

An adaptation to the real socio-economic situations appears necessary and may require a true paradigm shift in prioritizing R&D for companies or institutions. It should induce also a new approach of business models, combinations of product and service delivery or ways of addressing innovation processes and technologies.

A significant example is given with DNDi (Drug for Neglected Disease initiative) (Figs. 2 and 3). DNDi is a non-profit research and development organization the objectives of which are the delivery of new treatments for neglected diseases, more specifically sleeping sickness (human african trypanosomiasis), Chagas disease, leishmaniosis, specific (filarial) infections, but also malaria and paediatric HIV [2]. DNDi was launched in 2003 by Médecins Sans Frontières/Doctors Without Borders (MSF), the Oswaldo Cruz Foundation (FIOCRUZ) of Brazil, the Indian Council of Medical Research(ICMR), the Kenian Medical Research Institute (KEMRI), the Ministry of Health of Malaysia and the Pasteur Institute of France, the Special Program for Research and Training in Tropical Diseases (WHO-TDR) serving as permanent observer.

DNDi has already delivered six new treatments for neglected patients (Fig. 4): two fixed-dose combinations antimalarials (ASAQ and ASMQ), a late-stage sleeping sickness treatment with a combination of nifurtimox and eflornithine (NECT), a stibogluconate and paromomycin combination for visceral leishmaniosis

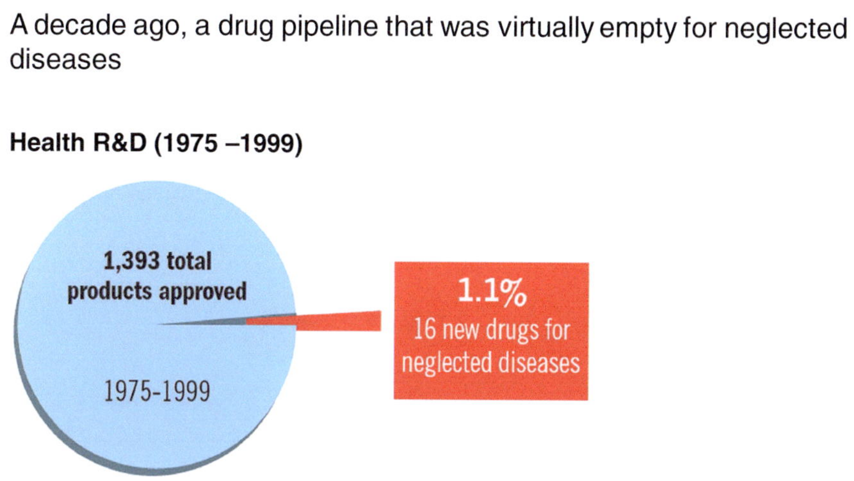

Fig. 2 The fatal imbalance. *Source* Fatal imbalance: The Crisis in Research and Development for Neglected Diseases, MSF, 2001

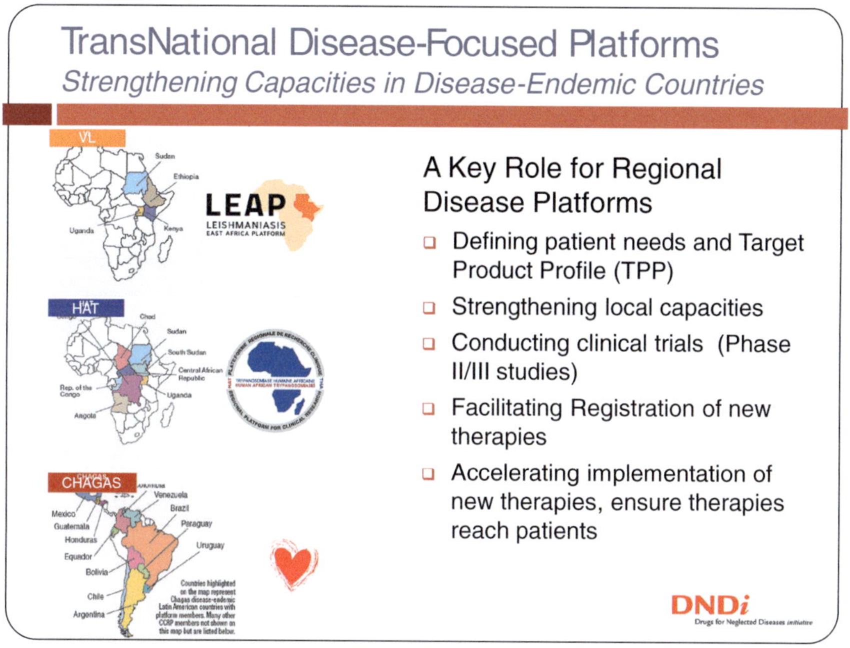

Fig. 3 DNDi R&D disease-focused platforms

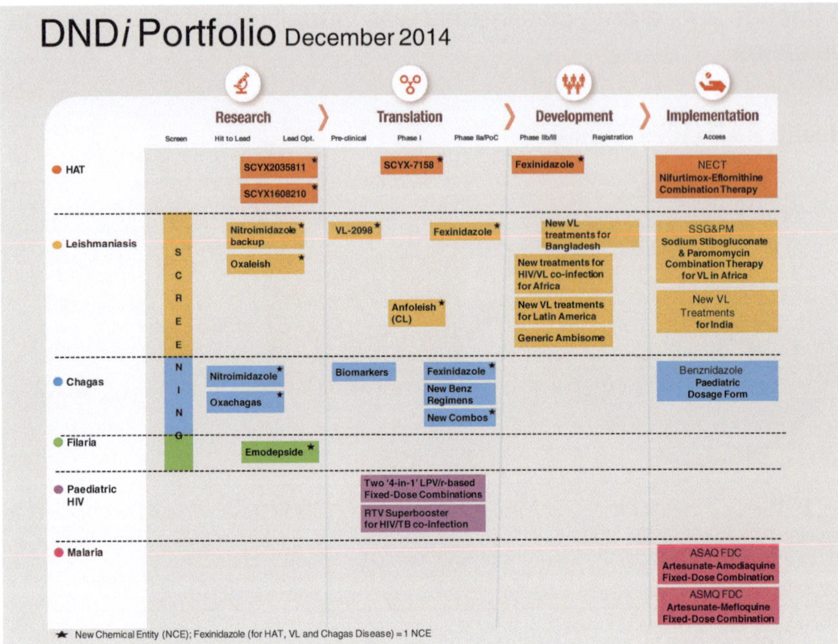

Fig. 4 DNDi portfolio

in Africa as well as various combination therapies for leishmaniosis in Asia, and a paediatric dosage form of benznidazole for Chagas disease in South America [2].

An estimated 3.4 million children have HIV/AIDS, but less than 28 % have access to antiretroviral therapy (ARV), compared with about 60 % for adults. Without treatment, more than half of these children will die before the age of two, and 80 % will die before the age of five [3]. Because of the absence of a viable market, major pharmaceutical companies have not invested in R&D specifically focused on the needs of young children with HIV/AIDS. The virtual elimination of mother-to-child transmission of HIV in high income countries means that nearly all HIV-positive children live in low-income countries and 90 % in sub-Saharan Africa. In spite of the global strategy to eliminate new infant infections through prevention of mother-to-child transmission (PMTCT) by 2015, a large children population will continue to be infected and will need access to early diagnostics and adapted ART combination therapies.

To fulfil the need for better paediatric formulations, DNDi is partnering with the Indian drug manufacturer Cipla Ltd. to develop two 4-in-1 ARV fixed-dose combination designed for children under the age of three, or until they are able to swallow tablets. These new combinations are made of tiny granules, which fit into a capsule that can be opened to incorporate the drugs onto soft food or be mixed with milk. The ARVs are «taste-masked», require no refrigeration and are easy to dose according to the child weight. Interactions with drugs for tuberculosis, a common co-infection with HIV have to be taken into account [3].

3 Reducing Time in the Hospital Through Efficient In-House Surveillance

Advanced systems enable remote patient monitoring or treatment and open the field of «connected health» and «e-health»: smart sensing technologies, as well as user interfaces, data processing and storage are important enablers beside smart phones and tablets in order to reduce healthcare expenses and improve patient compliance and interactions with medical staff.

The NATF has been working on the general concept of a fully integrated system called «Domomedicine» which includes all interventions and care delivered at home of patients or during their socio-professional activities. They should be at least equivalent in terms of quality and quantity compared to those provided in an hospital setting. Sometime complex, these activities require recent technologies with the objectives of maintenance of patients at home or at work with equivalent or better care, medical progress and reduced costs. The European project DOMOHEALTH is a user centered care model integrating healthcare and circadian rhythms. Starting from a strong base of related and successful previous or ongoing projects, DOMOHEALTH will develop and assess a novel concept of integrated care, based on advanced ICT systems and services, and recent research on «chrono-biology». DOMOHEALTH will be the first project to test an ICT-supported care model that will altogether:

- Address three chronic conditions (cancer, dementia and obesity), all with high European prevalence and a huge impact for older people. These conditions share co-morbidities and thus patient needs, and will be coordinated via a unique ICT platform based on Internet of Things, anchored in community and home care settings,
- Ensure patient privacy and data protecton to support digital collaborative multi-actor interaction and exchange of knowledge in respect of European Union (EU) and national regulations, thereby improving coordination and collaboration of care,
- Involve a wide range of local end-users, develop an extensive training and education program, targeting the various actors of health care chain as key success factors,
- Explore the monitoring and analysis of multimodal data on circadian performance as a basis for proactive patient-centered care, surpassing current reactive approaches and encouraging self-management,
- Provide a multi-dimensional proof of concept in field conditions, through multi-centre clinical trial involving 200 patients in 4 EU countries,

- Based on trials results, develop business models facilitating the deployment of DOMOHEALTH in the EU.

The DOMOHEALTH comprehensive approach is expected to provide a leap forward in the deployment of a new ICT-empowered care model, enabling patients and older persons to receive good care in their homes and reduce admissions into care institutions, optimizing quality of life whilst reducing the burden on healthcare systems. The consortium gathers leading experts from the ICT industry, clinical and multidisciplinary academic research, innovative Small and Medium Enterprises (SME), professional training and business planning.

4 Next Generation Sequencing (NGS) Technologies

The more recent generation of sequencing technologies has significantly reduced the cost of human genomic analysis and consequently raises a new question: how these technologies could help the communities to optimize the treatment of patients, enabling more personalized and targeted therapies? Besides these technologies

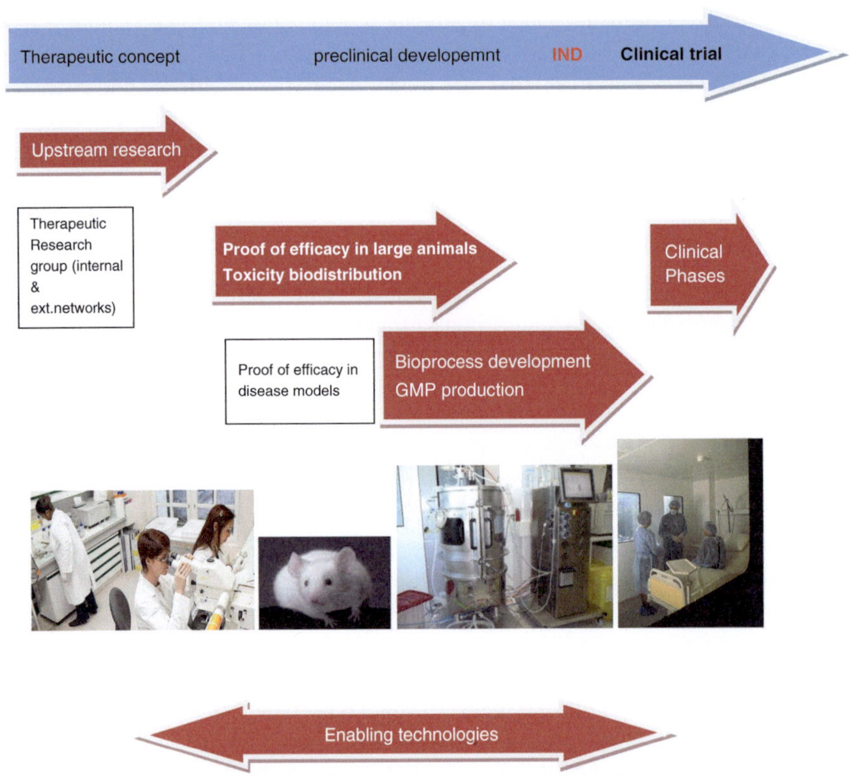

Fig. 5 Genethon: translational R&D for gene therapy

could typically improve the design of clinical development by a better selection of patients, and ultimately a better targeting of populations to be treated.

Treatment of rare genetic disease by gene or cell therapies has been the immediate consequence of genome sequencing. Recently gene therapy has been successfully used to treat several monogenic disorders. The first proof of concept was obtained in Severe Combined Immunodeficiency (SCID): the first trial for the treatment of X-SCID was performed by Pr. Alain Fisher and Pr Marina Cavazzana-Calvo using an ex vivo retroviral therapy [4]. This trial demonstrated the potential of this therapy and over 70 patients with SCID are now treated worldwide. In 2013 the European Medicines Agency approved the first gene therapy treatment for a monogenic deficiency, the lipoprotein lipase deficiency. Gene therapies could be considered for treating diseases with a more complex etiology, but for rare diseases with small number of patients, new business models will have to be invented [5]. Global consortia will facilitate the assembly of large cohorts of patients with specific rare diseases and patient organizations will have a key role in promoting these R&D activities. A good example is provided by Genethon, a gene therapy R&D facility in France created by the patient organization AFM-Telethon and dealing essentially with muscular distrophies, skin, eye, liver or immune monogenic diseases.

In spite of the low number of patients, production methods of gene therapy products are often a limiting factor and there is a real need for new processes with improved productivity and scalability. Focusing on process R&D, Genethon has opened one of the largest production facility worldwide for clinical grade gene therapy (Figs. 5 and 6).

Fig. 6 Genethon's production plant allowing production 20–30 GMP lots/year of AAV or LV depending on process

5 Conclusion

The rise of health-care costs remains a major issue in most countries even if the very large discrepancies which exist between developed or developing zones gives leeway for significant improvements in the prevention or treatment of acute or chronic diseases. From the few examples previously quoted, it appears that even in extreme situation like neglected or rare diseases, the best science and technologies bring efficient and economic response to health-care challenges. Consequently new technologies cannot be held accountable for all rising health-care costs but on the contrary they should play a major role in generalized access to better prevention and care. Some health economists are making the following statement that «the more health-care becomes digital, the more productive it will become». There are obvious physical and physiological limits to such assumptions but it has been clearly demonstrated that significant improvements are achievable even in the poorest or underprivileged environment, the key success factor being a total change of paradigm and a systemic approach of the objective.

One should also mention the large opportunities offered by the applications of Big Data in the field of health-care. Like in any other activities (production, marketing, finance, etc.) there are three dimensions to the data explosion that health organizations must control: volume (file size), velocity (the amount of data produced and handled at any given time) and variety (the types of data that must be handled and coordinated). Life sciences and health-care data have all these characteristics, particularly when considering the data produced by the next-generation sequencing (NGS) together with those associated with patient treatment and follow-up, data or image analysis, clinical interpretation or prevention. All these activities must be coordinated to get actionable decisions. Proving the efficacy of personalized or genomic medicines (which remain largely research driven) in terms of improved diagnostic, treatment regimens and costs will require not only an important investment in research and ICT but also a redefinition of the role of physicians as well as new hospital and health-care organizations.

References

1. *MIT Technology Review*, Business Reports, September 2013.
2. www.dndi.org.
3. *DNDi and Cipla Advance Development of Paediatric 4-in-1 ARVs to fullfil New WHO Guidelines*, Press release, 30 june 2013.
4. Salima Hacein-Bey-Abina, et al. N. Engl. J. Med., 2010, 363:355–364, July 22, 2010.
5. *Molecular therapy*, vol. 21 no 2 February 2013.

Big Data Analytics and Molecular Medicine

Kalyanasundaram Subramanian

Abstract Diagnostics have a major role to play in improving patient care, protecting consumer health and reducing health care costs. The quality of patient care can be significantly improved by detecting and diagnosing disease earlier and more rapidly. This is especially true in the case of cancer where accurate and early diagnosis can provide more targeted and effective treatment options leading to better outcomes. Diagnostic tests can also provide companies with accurate quality checks of their products thus ensuring product safety and consequently protecting consumer health.

Keywords Molecular diagnostics · Next generation genetic sequencing · NGS · Breast angiosarcomas · Molecular markers

1 Introduction

Diagnostics have a major role to play in improving patient care, protecting consumer health and reducing health care costs. The quality of patient care can be significantly improved by detecting and diagnosing disease earlier and more rapidly. This is especially true in the case of cancer where accurate and early diagnosis can provide more targeted and effective treatment options leading to better outcomes. Diagnostic tests can also provide companies with accurate quality checks of their products thus ensuring product safety and consequently protecting consumer health.

Classical diagnostic tests (e.g. CT scans, PET scans, MRIs, Serology tests, etc.) while commonly prescribed and used have several disadvantages. Imaging techniques such as CT scans, PET scans and X-rays require exposure to radiation while MRI is a lengthy and noisy procedure that cannot be performed in the presence of metal implants or pacemakers. Cellular diagnostics such as cell culture can have a

K. Subramanian (✉)
Strand Life Sciences, Pvt. Ltd., Bangalore, India
e-mail: kassub@gmail.com

R. Shorey and P. Ghosh (eds.), *Healthcare Engineering*,
DOI 10.1007/978-981-10-3111-3_6

long turnover period and the sensitivity depends on skill of the person performing the test as well as the condition of the sample. Additionally some viruses such as Hepatitis B, C and the parvovirus, cannot be grown in cell cultures. There are several disorders where the diagnosis is beyond the scope these methods, especially where the disease origin lies deep within the cell in the genome. To address these types of disorders, the field of molecular diagnostics has emerged where biological markers in the genome and/or proteome are identified to diagnose, monitor and detect risk of a disease to a patient [1]. This form of diagnosis is beginning to grow spawning the new field of personalized medicine [2]. There are various types of molecular markers and depending on the type, various techniques and approaches can be used. Broadly categorized they are genomic, proteomic, metabolomic and lipidomic.

2 Next Generation Sequencing

Sequencing seeks to determine the genetic code that underlies an individual by reading out the order and position of nucleotides in a gene. Several methods are used for sequencing with the Sanger sequencing method being the most widely used sequencing technology worldwide. However, it is expensive and time-consuming in cases where multiple genes need to be assessed. In the last 10 years, "next-generation sequencing" (NGS) technologies have made the process of sequencing high-throughput and relatively inexpensive compared with traditional methods [3].

Next generation sequencing (NGS) technologies encompass multiple methods and approaches that all have the commonality that they are high-throughput and massively parallel [4]. Over the past decade, their ubiquity in research has resulted in the Food and Drug Administration certification of the Illumina's MiSeqDx for CFTR diagnosis and the wide-spread use of several sequencers in LDTs—genome-based laboratory tests [5].

NGS applications include the identification of novel genomic variants by targeted re-sequencing specific genomic regions, genome wide scans to study variants, epigenetic markers and chromatin structures, describing the genomes of bacteria, plants and lower organisms by de novo assembly, studying the expression signatures of cells, tissues and organisms and studying hosts and species interactions by metagenomics [5].

3 How Does NGS Work?

NGS methodologies include library preparation, nucleic acid sequencing, imaging and data analysis. We first build a library of nucleic acids (gDNA or cDNA) by breaking the extracted DNA from tissue into small fragments of $\sim$500 bp and adding small, unique adapter sequences to the ends of the DNA fragments and amplifying them by emPCR or bridge amplification. The library is sequenced by the incorporation of known nucleotides in a sequentially and recording the newly formed sequences digitally by a fluorescence method. Each such sequence is called a read. This "raw" NGS data is analyzed first by computationally removing the adapter sequences and any poor-quality reads and then aligning the read-sequence to a known template sequence or in the case of previously unsequenced organism—bioinformatically reconstructing the sequence by de novo assembly. The compiled sequence may lend itself to several downstream processes and analyses such variant calling, novel gene or function identification, expression analysis, understanding of gene regulation and disease pathogenesis aided by several software tools both commercial in origin or from the public domain.

Understanding how the extent and position of DNA changes, whether small or large can focus ones attention on disease pathogenesis, diagnosis, risk prediction and treatment [6].

4 NGS Applications

NGS has enabled the relatively inexpensive and accurate detection of mutations both novel and rare for a large number of genes in a persons DNA efficiently. In rare inherited disorders, when there are multiple or unclear phenotypic manifestations, the simultaneous ability to analyze several mutated genes in a multi-gene panel by NGS improves the probability of a confirmed diagnosis [7].

In the area of cancer, this approach can help both identify risk mutations in families and also understand appropriate treatment options for patients based upon pathways implicated by their somatic mutational patterns [8]. When a cancer patient becomes unresponsive to therapy, NGS can aid a clinician in understanding the next steps, whether resistance mutations such as the p.Thr790Met mutation that makes a patient resistant to EGFR therapies, have evolved and if so, how the genetic signature can be used to plan therapeutic strategies to obviate the problem [9]. Hence it offers a dynamic view of tumor evolution enabling therapy to be tailored at each stage of the disease cycle.

In the following section we will demonstrate the utility of NGS in a variety of disease settings.

5 Case Studies

5.1 A Rare Inherited Disorder

A male child, born to a consanguineously married couple, showed diverse clinical manifestations including exophthalmos, recurrent vomiting, pulmonary and systemic hypertension, inguinal hernia, pyloric stenosis, pneumonia and secondary cardiac hypertrophy and died at the age of two of due to respiratory insufficiency. His elder sibling, also a male, manifested exophthalmos, pulmonary hypertension and respiratory problems, dying by year 1.

The patient's sample was subjected to whole exome sequencing. Two novel missense variants were identified, both of which could explain the observed clinical phenotypes—a homozygous variant, p.Met1Val, in the *FBLN5* gene, and a hemizygous variant, p.Asp1970Asn, in the *FLNA* gene. The *FLNA* gene encodes the cytoskeletal protein, filamin A, which plays an important role in the regulation of skeletal and brain development, blood vessel formation and clotting. Pathogenic variations in the *FLNA* gene are implicated in congenital heart and lung diseases including cardiac valvular dystrophy and pulmonary hypertension [10–12]. The identified *FLNA* variant was mapped to the R18 domain, which is critical for the interaction of the protein with several binding partners. Loss of diverse protein interactions could potentially result in the phenotypically heterogenous clinical manifestations seen in this patient.

Sequencing of the parents' DNA revealed that the father did not harbor the variant and the mother had one bad copy of the gene (heterozygous). Sons born to this apparently healthy couple are expected to be affected, while daughters are likely to be sub-clinical. When the parents were pregnant with a third child, prenatal diagnosis of the embryonic cells revealed that the foetus was female and predicted to be healthy. This was what was observed when a healthy daughter was delivered.

5.2 Treating a Rare and Deadly Cancer

Breast angiosarcomas represent ~8 % of all angiosarcomas which in turn constitute a set of rare malignancies [13] that are aggressive with poor prognosis [14].

The patient was a 24 year old woman diagnosed with bilateral metastatic multicentric breast angiosarcoma (largest Rt—6 × 5 × 4 and largest left—9.8 × 7 × 5). Post cytoreductive breast surgery, 4 cycles of iphosphamide/doxorubicin-based adjuvant chemotherapy were administered to the patient. Follow up with PET-CT scan demonstrated hypermetabolic left breast lesions and weakly metabolic right breast nodule, thus indicating presence of residual viable disease. Another round of follow up with an ultrasound showed reduction in number of nodules and reduction in size but still amounting to presence of disease.

To explore targeted therapy options, the patient tumor sample was sequenced for somatic mutations in 48 cancer relevant genes. The test revealed a somatic mutation in *PIK3CA*, a gene that is involved in driving cellular growth and proliferation through the PI3K/AKT/mTOR pathway. Activating mutations in *PIK3CA* are oncogenic as they result in growth factor-independent cell proliferation and protection from cell death (Samuels et al.). Clinical and preclinical studies show that tumors with an active PI3K/AKT/mTOR pathway may be treated with drugs that block or suppress the pathway [15]. Anti-mTOR drugs such as everolimus and temsirolimus are approved by FDA for use in other cancers show in anti-tumor activity in preclinical studies in angiosarcoma cell lines [16].

Everolimus was selected and administered to the patient at 10 mg/day. A three month follow up with ultrasound showed the complete absence of any solid or cystic mass lesion in the breast. NGS based mutation profiling expanded the therapy options for the patient and allowed the discovery of a potential cure.

6 Conclusions

NGS combines informatics with biology to bring the promise of precision medicine to a clinical reality. It is remarkable that in less than 15 years since the Human Genome Project, the entire human genome can be sequenced within a single day. This evolution in sequencing has enabled personalized genomic medicine, allowing clinicians to identify with high confidence, several different types of genomic variations. One can witness the exponentially growing usage of NGS methodologies in carrier testing, prenatal detection of fetal aneuploidies, diagnosis of rare diseases, estimation of familial cancer risk and cancer treatment. The principal challenge in the future is to reduce costs further and make NGS-based genetic tests more accessible to patients thus ensuring that the significant advances made in this field are put to best use.

References

1. Poste, George. "Molecular diagnostics: a powerful new component of the healthcare value chain." *Expert review of molecular diagnostics* 1.1 (2001):1–5.
2. Hamburg, Margaret A., and Francis S. Collins. "The path to personalized medicine." *New England Journal of Medicine* 363.4 (2010): 301–304.
3. Precone V et al. 2015. Cracking the Code of Human Diseases Using Next-Generation Sequencing: Applications, Challenges, and Perspectives. Biomed Res Int. 2015:161648.
4. Luthra R et al. 2015. Next-Generation Sequencing in Clinical Molecular Diagnostics of Cancer: Advantages and Challenges. Cancers (Basel). 7(4):2023–2036.
5. Bahassi el M et al. 2014. Next-generation sequencing technologies: breaking the sound barrier of human genetics. Mutagenesis. 29(5):303–310.
6. Jiang T et al. 2014. Application of next-generation sequencing technologies in Neurology. Ann Transl Med. 2(12):125.
7. Madrigal I, et al. 2014. Efficient application of next-generation sequencing for the diagnosis of rare genetic syndromes. J Clin Pathol. 67(12):1099–1103.
8. Yan-Fang Guan, et al. 2012. Application of next-generation sequencing in clinical oncology to advance personalized treatment of cancer. Chin J Cancer. 31(10): 463–470.
9. Jänne PA et al., 2015. Engl J Med. 372 (18):1689–1699.
10. Masurel-Paulet A et al. 2011. Lung disease associated with periventricular nodular heterotopia and an FLNA mutation. European journal of medical genetics. 54(1):25–28.
11. Nakamura F et al. 2011. The filamins: organizers of cell structure and function. Cell adhesion and migration. 5(2):160–169.

12. Ruskamo S et al. 2012. The C-terminal rod 2 fragment of filamin A forms a compact structure that can be extended. The Biochemical journal. 446(2):261–269.
13. Fury MG et al., (2005) A 14 year retrospective review of angiosarcoma: clinical characteristics, prognostic factors, and treatment outcomes with surgery and chemotherapy. Cancer J. May-Jun;11(3):241–247.
14. Hu QC et al., (2015) Management experiences of primary angiosarcoma of breast: a retrospective study from single institute in the People's Republic of China. Onco Targets Ther. 8:3237–3243.
15. Gonzalez-Angulo AM et al., (2013) Defining biomarkers to predict sensitivity to PI3K/Akt/mTOR pathway inhibitors in breast cancer. Cancer Treat Rev. 39(4):313–320.
16. Jiang BH et al., (2000) Phosphatidylinositol 3-kinase signaling mediates angiogenesis and expression of vascular endothelial growth factor in endothelial cells. Proc Natl Acad Sci U S A. 97(4):1749–1753.
17. Samuels Y and Waldman T (2010) Oncogenic mutations of PIK3CA in human cancers. Curr Top Microbiol Immunol. 347:21–41.
18. Young RJ et al., (2014) Vascular-targeted agents for the treatment of angiosarcoma. Cancer Chemother Pharmacol. 73(2):259–270.

Can Life Sciences Progress Without Engineering?

Sanghamitra Bandyopadhyay

Abstract Life science has traditionally been driven by wet lab experiments carried out in the laboratory, with important observations emerging from these efforts. However, since early days itself, life sciences had to heavily depend on discoveries in physics, chemistry and engineering for making its inferences. Starting from the first microscope developed in the early seventeenth century, technology has developed in leaps and bounds, making it possible now to investigate human genome of the individuals within a day. Over the past few decades, major developments in the field of molecular biology coupled with advances in genomic technologies have led to an explosive growth in the biological information generated by the scientific community. Every aspect of life is now inextricably touched by engineering and technological developments, providing a natural answer to the question raised in the title—No, life sciences can no longer do survive without engineering. In this article, we touch upon some of such important developments that have forever changed the way biological research will be conducted.

Keywords Interdisciplinary research · Bioengineering · Molecular biology · Bioinformatics · Computational biology · Systems biology

1 Introduction

We can envisage the future of life sciences from a recent MIT report [1] that calls it the third revolution in biological sciences—the convergence of life sciences, physical sciences and engineering. While the developments in cellular and molecular biology has guided the first revolution and the developments in genomics has driven the second revolution, convergence of the different disciplines is what provides a paradigm shift in the way research will be conducted in future [1].

S. Bandyopadhyay (✉)
Indian Statistical Institute, 203 BT Road, Kolkata 700108, India
e-mail: sanghami@isical.ac.in

R. Shorey and P. Ghosh (eds.), *Healthcare Engineering*,
DOI 10.1007/978-981-10-3111-3_7

The most fundamental discovery in molecular biology was the understanding of the central dogma, and the double helical structure of the DNA [2]. The central dogma provided the basic blueprint of how proteins are produced from DNA—the processes of transcription that makes the RNA from the DNA, followed by translation that produces the proteins from the RNA.

Decoding the genome by figuring out the sequence of characters in the DNA, comprised the most important discovery in the era of genomics. The human genome project, which took 13 years after getting started in 1990, was an international effort that saw the involvement of different countries around the globe. The two revolutions together provided insights into the working of the cell at levels previously unheard of. Technology developed at an astounding pace, leading to the generation of biological data of unimaginable magnitude. Together with these amazing developments came numerous challenges. Challenges of developing computational methods and statistical models for analyzing this data in order to better understand the cellular processes and make newer discoveries. Computational biology and bioinformatics, the areas that have evolved in response to these challenges, can be viewed as the *use of computational methods to handle biological data.*

2 Basics of Molecular Biology

Deoxyribonucleic acid (DNA) and *proteins* are biological macromolecules built as long chains of chemical components, namely nucleotides and amino acids, respectively. Proteins are compounds whose production templates exist in specific regions of the DNA that are called genes. Through such genes, the hereditary information gets transmitted from one generation to the other. Each nucleotide making up the DNA contains a base molecule that can be one of two types, purine (*adenine* (A) and *guanine* (G)) and pyrimidines (*cytosine* (C) and *thymine* (T)). In addition, a nucleotide contains a sugar (ribose in RNA and deoxyribose in DNA), and one or more phosphate groups. The DNA forms a double helix of two complementary strands bonded together, with hydrogen bonds holding them together. It may be noted that besides genes, the DNA strand encodes for many other molecules, typically different types of RNA. In fact, it is well known that only about 3–5 % of the DNA sequence codes for proteins, the remaining portion traditionally thought to be junk. However, recent studies point to the fact that these so-called junk regions contain very specific regulatory information.

The genes undergo a transformation process called transcription to form messenger RNAs, which thereafter undergo translation to form proteins. These two processes together are commonly referred to as the *central dogma* of molecular biology. Proteins are linear chains of amino acids which bond with each other by peptide bonds eliminating water molecules in the process. Twenty standard *amino acids* (or "residues") are known, each of which is coded by one or more triplets (or *codons*) of the nucleotides making up the DNA. Based on the *genetic code*, the

DNA is transformed into a protein. Proteins are involved in almost every cellular activity with myriad functions such as molecular transport, creating ionic channel, catalyzing biochemical reaction.

3 Bioinformatics, Computational Biology and Systems Biology

Bioinformatics is viewed as the exploitation of computational methods to make biological discoveries. It is sometimes synonymously used with computational biology. However, some researchers tend to ascribe subtle differences between the two. Whereas bioinformatics mostly deals with applying tools and methods directly on the biological data, computational biology looks at models to understand the functionalities of biomolecules. Systems biology, on the other hand, looks at an organism as a system consisting of interacting actors, forming myriad of complex networks at different levels.

Discovering new biological insights and deriving unifying biological and biophysical principles is one of the primary objectives of research in computational biology. The final result of such endeavors will be a better understanding of the biological system, thereby leading to methods and interventions for improvement of the standard of living. Sequence analysis, phylogenetic/evolutionary trees, protein classification and analysis of microarray data constitute some typical problems of bioinformatics where mining techniques are required for extracting meaningful patterns. The mining tasks often used for analyzing biological data include clustering, classification, prediction and frequent pattern identification, etc. The huge amount of biological data stored in repositories distributed all around the globe is often noisy. Moreover, the same information may be stored in different forms. Therefore, data preprocessing tasks such as cleaning and integration are important in this domain. Clustering and classification of gene-expression profiles or microarray data are performed in order to identify the genes that may be responsible for a particular trait. Determining or modeling the evolutionary history of a set of species from genomic DNA or amino acid sequences using phylogenetic trees is widely studied in bioinformatics. Classification of proteins and homology modeling are two important approaches for predicting the structure of proteins, and may be useful in drug design. Motif-based classification of proteins is also another important research direction. A motif is a conserved element of a protein sequence that usually correlates with a particular function. Motifs are identified from a local multiple sequence alignment of proteins corresponding to a region whose function or structure is known. Motif identification from a number of protein sequences is another mining task that is important in bioinformatics. Similarly, it is also important to explore various structural and network motifs for various other applications.

4 Important Tasks in Computational Biology

Most of the tasks in computational biology are broadly grouped into four categories—those dealing with sequences, structures and expression values, and finally those at the systems level. However, with the recent progress of genomewide epigenetic analysis another dimension has been added to this [3].

As already mentioned DNA, RNA and proteins may all be viewed as molecular sequences. Some of the important sequence level tasks include fragment assembly, sequence alignment, gene finding, phylogenetic tree construction and sequence classification.

Prediction of 3D structure from 1D sequences, protein folding, structure based protein classification, molecule design and docking, etc. constitute structure level tasks.

Microarray expression datasets are frequently analyzed for a variety of predictions. These include finding differentially expressed genes, gene clustering, classification of gene expression data, differential co-expression analysis, gene ordering, etc.

Although it is important to understand the cellular components at an individual level, the cell is actually a combination of its components that interact with each other in complex ways. A biological system is more than its components, and hence variability is high although the system is robust. Therefore systems level analysis is of great importance [4]. Such tasks include understanding the dynamics of intra and intercellular processes that determine cell function, gene regulatory networks and metabolic pathways. Related tasks in this category are study of survival prediction, cancer prediction, drug response, drug discovery, drug delivery, schedule optimization, etc.

5 Engineering and Computational Challenges in Biology

Interdisciplinary studies have now become the staple of biological research. While medical devices constitute an obvious application of engineering to biology, the technological and computational advancements have also made significant impact. One of the earlier computational challenge faced was in sequencing the genome. Here, the technology on offer could only sequence shorter fragments of the DNA. Reconstructing the full genome posed a grave problem, which was solved by using multiple copies of the same genome, with an elegant computational solution referred to as the shotgun sequencing. This classical approach of sequencing was referred to as the Sanger sequencing method. Since the last 10 years, more sophisticated technologies have been developed that produce millions of shorter reads resulting in more coverage of the genome. However, shorter reads give rise to computational challenges in mapping them to a reference genome, since these are associated with the problem of matches at multiple locations.

On the different side, several successful efforts have been carried out to achieve a better understanding of diseases like the EURESIST programme by European Commission targeting AIDS and OSDD programme by Council of Scientific and Industrial Research, India for TB and Malaria. This highlights the requirement of large-scale multi-disciplinary attempts to assist biology.

There are several computational challenges involved in solving the bioinformatics problems. First of all, biological datasets, such as microarray or next generation sequencing data are usually huge, especially due to the popularity of population genomics. Therefore, the algorithms for managing and analyzing such huge amount of data need to be computationally efficient. On the other hand, many experiments involve lack of data, such as non-coding RNAs, and thus the computational methods should be able to handle this less amount data in a robust way. Furthermore, biological datasets are usually noisy and also there may be missing values. So, good preprocessing techniques are necessary for these datasets before applying the actual algorithms. The recent progress in bioinformatics and computational biology must deal with these kinds of problems.

Here, we cover some applications and recent advances in different bioinformatics domains. This includes gene and microRNA expression data analysis using clustering, classification and feature selection, rational drug design, microRNA target prediction, gene-transcription factor-microRNA network building and analysis, intra and inter-species protein-protein interaction prediction and analysis, algorithms for next generation sequencing etc. Use of computational intelligence and other tools for solving these problems is extremely appealing.

Last but not the least is the recent progress of genome editing technologies that facilitate the insertion, deletion and replacement of a DNA segment [5]. There are enormous scopes in fighting against diseases if we can cope up with these engineering advances to be applicable in life sciences research.

6 Conclusion

Bioinformatics and computational biology are the areas of research that skillfully apply computational methods for solving biological problems. Here we discuss the fundamentals of molecular biology followed by different facets of bioinformatics and computational biology. Moreover, we emphasize on the challenges involved in applying computational methods in bioinformatics domain. Finally, some applications and recent advancements in the field of bioinformatics and computational biology are discussed. Overall, this provides a brief insight into the past, present and future synergy between life sciences and engineering.

References

1. P. A. Sharp, S. N. Bhatia, C. L. Cooney, T. E. Jacks, M. A. Kastner, D. A. Lauffenburger, J. Lees, R. Langer, R. Sasisekharan, P. T. Hammond, M. B. Yaffe, M. Sur, The Third Revolution: The Convergence of the Life Sciences, Physical Sciences, and Engineering, MIT White Paper, 2011 (Link: http://dc.mit.edu/sites/dc.mit.edu/files/MIT%20White%20Paper%20on%20Convergence.pdf).
2. F. Crick, Central dogma of molecular biology, Nature, 227(5258):561–563, 1970 (DOI:10.1038/227561a0).
3. A. P. Feinberg, Epigenetics at the epicenter of modern medicine, The Journal of the American Medical Association, 299(11):1345–50, 2008 (DOI:10.1001/jama.299.11.1345).
4. U. Sauer, M. Heinemann, N. Zamboni, Genetics: Getting Closer to the Whole Picture, Science, 316(5824):550–551, 2007 (DOI:10.1126/science.1142502).
5. A. Petherick, Genome editing, Nature, 528:S1, 2015 (DOI:10.1038/528S1a).

Evaluation of BMP-2 Mediated Bone Formation Using Enzymatically Crosslinkable Injectable Hydrogels: An In Vivo Study Using Transgenic Fluorescent Reporter Mouse Model

Shalini V. Gohil and Lakshmi S. Nair

Abstract The study highlights the potential of Col3.6 transgenic fluorescent reporter mouse models to understand BMP mediated localized bone formation, using enzymatically crosslinkable glycol chitosan hydrogel as a delivery vehicle. Using the injectable gel, rhBMP-2 (0.5 μg/defect) was able to fully regenerate the critical sized calvarial defect in vivo at 8 weeks. At a cellular level, the tissue showed the presence of mature osteoblasts (bright green) with active mineralization (AC) and osteogenic activity (ALP) as well as presence of osteoclasts (TRAP) carrying out active remodeling, indicating the quality and maturation of the mineralized, regenerated tissue.

Keywords Injectable hydrogel · Glycol chitosan · Transgenic reporter mouse · Calvarial bone defect model

1 Introduction

There is a significant clinical need to develop injectable delivery vehicles that can retain and localize rhBMP-2 mediated bone formation at the implantation site. Such delivery vehicles could decrease the dose and associated side effects of rhBMP-2,

S.V. Gohil · L.S. Nair (✉)
Department of Orthopedic Surgery, UConn Health, Farmington, CT 06030, USA
e-mail: nair@uchc.edu

S.V. Gohil · L.S. Nair
Institute for Regenerative Engineering, The Raymond Beverly Sackler Center for Biomedical, Biological, Physical and Engineering Sciences, UConn Health, Farmington, CT 06030, USA

L.S. Nair
Departments of Material Science and Engineering, Biomedical Engineering and Institute of Material Science, University of Connecticut, Storrs, CT 06269, USA

R. Shorey and P. Ghosh (eds.), *Healthcare Engineering*,
DOI 10.1007/978-981-10-3111-3_8

increase the cost-effectiveness and also present an option to deliver rhBMP-2 in a minimally invasive manner. A challenge for the tissue regeneration field lies in understanding the cellular basis for an experimental outcome. Transgenic fluorescent reporter models present unique advantages to understand the cellular contribution in rhBMP-2 mediated localized bone formation in the hydrogel [1–3]. Our previous study in a bilateral calvarial defect model showed that the injectable hydrogels prepared by enzymatic crosslinking of 3-(4-hydroxyphenyl)propionic acid-modified glycol chitosan (HPP-GC) show good spatial control on rhBMP-2 mediated bone formation, even when high dose (2 μg/defect) of rhBMP-2 is used [1]. The present study aims to analyze the bone formation capability of these injectable HPP-GC hydrogels at lower rhBMP-2 dose (0.5 μg/defect) and to evaluate the quality of bone formation at a cellular level using transgenic fluorescent reporter mouse model.

2 Methods

A bilateral, critical sized bone defect model was developed in CD-1 transgenic mice containing 3.6-kb fragment of the rat collagen type 1 promoter fused to topaz fluorescent protein (Col3.6Tpz; EYFP, green), as described earlier [1]. The defects were implanted with HPP-GC hydrogel loaded with 0.5 μg rhBMP-2 (HPP-GC + 0.5 μg BMP), for 8 weeks. One day before sacrifice, the mice were injected alizarin complexone (30 mg/kg, IP). After sacrifice, calvaria were dissected, fixed in 10 % formalin, and imaged by xray and microCT. The samples were then fixed in Shandon Cryomatrix^{TM}, cryosections (5 μm) were obtained and sequentially imaged by darkfield imaging, alkaline phosphatase (ALP) staining and tartrate-resistant acid phosphatase (TRAP) staining [3].

3 Results and Discussion

The biological function of rhBMP-2 is to recruit and activate participating osteoprogenitor cells to form bone. Current translational research is focused on understanding how rhBMP-2 needs to be presented at the regeneration site to achieve this biological function effectively. Figure 1a shows the feasibility of HPP-GC to locally deliver rhBMP-2 at the defect site in a biologically active manner resulting in complete regeneration of critical size calvarial defect at a concentration much lower than that reported before [1]. The spatially confined complete bone regeneration observed in the present study demonstrates the ability of the enzymatic crosslinking process to retain the biological activity of rhBMP-2 and localize the activity at the site of implantation [1]. Even at this lower concentration, the retained protein in the hydrogel was able to recruit and activate osteoprogenitor cells as evidenced from this study using transgenic Col3.6 topaz fluorescent reporter mice. The osteoprogenitor cells in Col3.6 topaz mice express mild fluorescence in preosteoblasts and exhibit

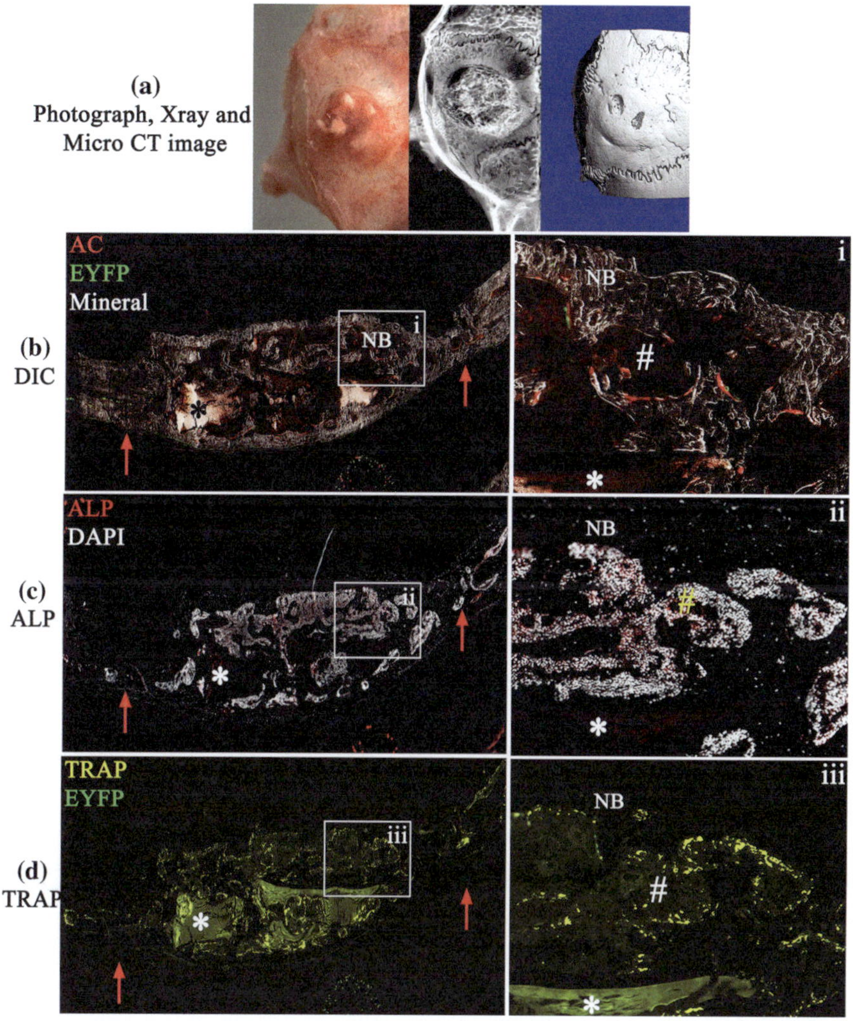

Fig. 1 New bone regeneration in HPP-GC+0.5 μg rhBMP-2 implanted critical sized defects at 8 weeks showing **a** Photograph, X-ray and MicroCT images; **b** Darkfield image showing EYFP positive (*green*), AC (*red*) and Mineral (*white*) signals; **c** TRAP staining showing TRAP positive (*yellow*) cells; **d** ALP staining showing ALP positive (*red*) cells with cell nuclei (*white*). ***i–iii*** show the magnifications of boxed areas (*) Non-degraded gel, *NB* New bone; # Marrow like matrix; *Red arrows* indicate defect edges

strong green fluorescence in mature osteoblasts [4, 5]. Actively mineralizing osteoblasts show a sharp red mineralization line upon AC injection. ALP and TRAP staining can be used to confirm the presence of osteoblasts and remodeling osteoclasts. In the present study, the DIC imaging (Fig. 1b) of the histological section of non-decalcified, regenerated tissue at the defect site (red arrows indicate the defect edges) showed the presence of bright green osteoblasts with underlying red AC label in the new bone (NB) (Fig. 1b i). These areas showed co-localization of ALP activity (Fig. 1c, ii) further confirming the osteogenic activity of and active mineralization by these osteoblasts. The sections also showed presence of non-degraded hydrogel (*) and marrow-like areas (#). The presence of TRAP positive osteoclasts (Fig. 1d, iii) indicated active remodeling of the regenerated matrix, which is necessary for its progression into a mature and functional bony tissue. The presence of new bone, fluorescent green osteoblast cells with overlying red AC label and TRAP stain indicating osteoclast activity clearly demonstrates the ability of the gel to support balanced and active mineralization and remodeling.

4 Conclusions

This study demonstrates the ability of HPP-GC hydrogels to retain the biological activity of rhBMP-2 at lower concentrations and support localized bone formation. The transgenic fluorescent reporter mice can serve as a functional read out of the active bone formation and remodeling process at a cellular level to understand tissue regeneration using regenerative engineering strategies.

References

1. Gohil SV, Adams DJ, Maye P, Rowe DW, Nair LS. Evaluation of rhBMP-2 and bone marrow derived stromal cell mediated bone regeneration using transgenic fluorescent protein reporter mice. J Biomed Mater Res A. 2014;102(12):4568–80.
2. Gohil SV, Brittain SB, Kan H-M, Drissi H, Rowe DW, Nair LS. Evaluation of enzymatically crosslinked injectable glycol chitosan hydrogel. J Mater Chem B. 2015;3:5511–22.
3. Gohil SV, Bagshaw KR, Rowe DW, Nair LS, editors. Injectable chitosan hydrogels with tailored degradation and release properties for localized biomolecule delivery Society for Biomaterials Annual Meeting and Exposition, April, 2014; 2014; Denver, Colorado.
4. Jiang X, Kalajzic Z, Maye P, Braut A, Bellizzi J, Mina M, et al. Histological analysis of GFP expression in murine bone. The journal of histochemistry and cytochemistry : official journal of the Histochemistry Society. 2005;53(5):593–602.
5. Boban I, Jacquin C, Prior K, Barisic-Dujmovic T, Maye P, Clark SH, et al. The 3.6 kb DNA fragment from the rat Col1a1 gene promoter drives the expression of genes in both osteoblast and osteoclast lineage cells. Bone. 2006;39(6):1302–12.
6. Dacic S, Kalajzic I, Visnjic D, Lichtler AC, Rowe DW. Col1a1-driven transgenic markers of osteoblast lineage progression. J Bone Miner Res. 2001;16(7):1228–36.

Supporting the Diagnosis of Childhood Pneumonia in Low Resource Settings

Pavan Dadlani, Ercan Gigi, Ernest Laman, Eefje Aarts and Helle Ullerup

Abstract Pneumonia is the leading infectious disease killer of children under 5 worldwide [1]. 99 % of deaths occur in developing countries in low-resource settings [2]. The World Health Organization (WHO) has implemented Community Case Management (CCM) in rural areas to help diagnose and treat the main diseases in children under 5. CCM empowers members of the community, known as Community Health Workers (CHWs), to identify danger signs of pneumonia, a.o. counting the number of breaths per minute and comparing the result with defined cut-off rates per age group.

Keywords Pneumonia · Low-resource settings · Diagnostics · Respiratory rate

1 Introduction

Pneumonia is the leading infectious disease killer of children under 5 worldwide [1]. 99 % of deaths occur in developing countries in low-resource settings [2]. The World Health Organization (WHO) has implemented Community Case Management (CCM) in rural areas to help diagnose and treat the main diseases in children under 5. CCM empowers members of the community, known as Community Health Workers (CHWs), to identify danger signs of pneumonia, a.o. counting the number of breaths per minute and comparing the result with defined cut-off rates per age group. Measuring breathing rate is the most challenging task for CHWs and current tools (e.g. timers) are not easy to use, are very distracting, and often lead to miscounting the breaths. Insufficient education and technology leads to poor diagnosis, resistant bacterial cultures due to overtreatment, higher costs of care and unnecessary deaths.

P. Dadlani (✉) · E. Gigi · E. Laman · E. Aarts · H. Ullerup
Philips Research, Eindhoven, The Netherlands
e-mail: pavan.dadlani@philips.com

R. Shorey and P. Ghosh (eds.), *Healthcare Engineering*,
DOI 10.1007/978-981-10-3111-3_9

2 Methods

We developed ChARM (Children's Automated Respiration Monitor) in response to UNICEF's call for innovation for improved diagnostics. ChARM is placed around the child's belly and automatically measures respiratory rate. An algorithm converts chest movements detected by a 3D accelerometer into breath counts, and classifies fast breathing according to WHO guidelines.

To develop and test the algorithms, accelerometer-based respiration data was collected from children (0–5) in different postures. To test the accuracy of the device, we used the respiratory rate computed by manually annotated video recordings of each child as a reference, and compared it to the ChARM's measurement.

ChARM's user interface and feedback is designed for low-skilled CHWs. The device went through an iterative design process by actively engaging with the Global Health Community and conducting field tests in East Africa and India. Over 100 health workers gave feedback in focus group and interviews, ranging from different levels of care and covering rural and urban workers.

3 Methods

ChARM has demonstrated acceptable levels of accuracy (root mean squared error (RMSE) of within 2 breaths). The device has been highly appreciated and endorsed by the Global Health Community. Health workers valued the on-body measurement while providing quantitative and qualitative feedback in an intuitive way. It would radiate trust and would help them communicate the results to the parents.

4 Conclusions

ChARM has the potential to assist CHWs to make more accurate measurements of a child's breathing rate and improve the diagnosis of pneumonia. This can ensure the appropriate prescription of antibiotics, but also potentially preventing many of the deaths caused by pneumonia.

References

1. http://www.worldpneumoniaday.org/.
2. UNICEF. 2014. "Country estimates of child mortality, causes of under–ve deaths, and coverage indicators" in Committing to Child Survival: A Promise Renewed, Progress Report. http://-les.unicef.org/publications/-les/APR_2014_web_15Sept14.pdf.

Adoption of Personalized Medicine: Towards Identifying Critical Changes

Subhas Chandra Misra and Sandip Bisui

Abstract Among the emerging areas in the health-care system, the implementation of Electronic Medical Record system and the discovery of Personalized Medicine are quite prominent. While scientists have already discovered some of the personalized drugs, implementing this new Medicare system requires a lot of changes in the traditional health-care system. This paper aims at identifying these critical changes required in the adoption of the Personalized Medicine systems.

Keywords Personal medicine · Medicare · Healthcare costs · Likert scale · ICT-based medicine

1 Introduction

Among the emerging areas in the health-care system, the implementation of Electronic Medical Record system and the discovery of Personalized Medicine are quite prominent. While scientists have already discovered some of the personalized drugs, implementing this new Medicare system requires a lot of changes in the traditional health-care system. This paper aims at identifying these critical changes required in the adoption of the Personalized Medicine systems.

S.C. Misra (✉) · S. Bisui
Industrial and Management Engineering,
Indian Institute of Technology (IIT) Kanpur, Kanpur, India
e-mail: subhasm@iitk.ac.in

R. Shorey and P. Ghosh (eds.), *Healthcare Engineering*,
DOI 10.1007/978-981-10-3111-3_10

2 Methodology

We undertook a survey based ex-post facto empirical study approach. First, we formulated the problem and identified a possible list of possible changes required for the adoption based on previous literature available. Then we designed a closed ended survey questionnaire on 5-point Likert scale for medical doctors/medical researchers and patient groups.

3 Results

From the analysis of the data obtained, eleven factors out of fourteen factors turn out to have significant impact in the change management of the adoption of Personalized Medicine. These are: from reactive to efficient medical care, from trial-and-error to right treatment for right person at right time, changes associated with PM specific training for medical practitioners, from narrow mind-set to open minded-ness of patients' community, from open information of patients to secure information of patients, from less emphasis on IT infrastructure to more emphasis on IT infrastructure, from resistance to change to openness to change, changes in technology, individual profit based drugs production to collaborative personalized drugs production, non-accessibility of patients' information to accessibility of patients' information and from small scale to large scale data and knowledge management.

Based on the existing literature and the intuitive analysis the impact of these factors on the change management of Personalized Medicine is strongly justifiable. Further research could be done by considering larger population consisting of both urban and rural people. In the next phase, we also have planned to take the opinion of medical researchers, practitioners in pharmaceutical companies, the regulatory board members as well as patients, and analyze the perspectives of all these categories of respondents. This will be very helpful in implementing Personalized Medicine systems in the years to come.

4 Conclusion

We have statistically analysed and identified a set of possible changes required for the adoption of Personalized Medicine. This research result will enable to envision the e-health systems.

Author Biographies

Subhas Chandra Misra is currently an Associate Professor at IIT Kanpur and an Advisor at Indo Canada Education Council. Dr. Misra has experience in both academia and industry. Before joining IIT Kanpur, he had worked for several years in different industries including Indian Telephone Industries Ltd (India) and Nortel Networks (Canada). He received his PhD degree from Carleton University, Ottawa, Canada and a PDF from Harvard University, Boston, USA. His research interests include, Healthcare Management, Personalized Medicine, Business Process Management, Project Management, Managing Supply Chain Services, Business Analytics/Data Analysis, Enterprise Resource Planning (ERP), Change and Risk Management, E-Governance, and Information Systems.

He has authored more than 80 peer reviewed papers which have received high citation index. His papers have been published in reputed international journals (of high impact factor) published by extremely reputed publishers like Elsevier, Springer, Wiley, IEEE and Emerald. He has received 20 prestigious awards for his academic recognitions that include "Research Excellence Award" (USA), "Achievement Award" (USA), Best/Outstanding Paper Award (USA), Best paper Award (Canada), NSERC Fellowship (Canada), OGS (Canada), Outstanding Faculty Award, and Young Scientist Award. He so far was invited as Keynote Speaker in 20 international conferences in North America, Europe and Asia including the most prestigious World Congresses held in the venues: Imperial College (London, UK) and UC Berkeley (California, USA), where his novel concepts were taken very warmly. He has served as Associate Editor of 12 prestigious international journals published across the globe by top international publishers. He has been invited as Program Chair and Session Chair in TWELVE international conferences held around the globe (Europe and USA). He has served as conference programme committee member for more than 50 International Conferences held across USA, Canada and EUROPE.

Sandip Bisui is currently pursuing his B.S-M.S Dual Degree at Department of Mathematics and Statistics, Indian Institute of Technology (IIT) Kanpur. His research interest lies in **Healthcare Management**, **E-Health**, **EMR, Personalized Medicine, ERP** and **Statistical Modelling.** Along with **Dr. Subhas Chandra Misra** [PhD (Carleton), PDF (Harvard); ASSOCIATE PROFESSOR, IME Department, IIT Kanpur], he has published a couple of papers in the area of Personalized Medicine. Currently he is working on several projects related to e-health with Dr. Subhas Chandra Misra.

Biophysics-Based Markers Surpassing Biochemical Markers in Screening of Ageing-Related Neurodegeneration and Cognitive Impairment

Rajiv Ramaswamy, Rajesh Khanna and Prasun Roy

Abstract Historically the first biomarker to distinguish abnormal from normal tissue was biophysical properties, such as elasticity of tissue or viscosity of body fluids. Based on the stiffness or percussion of organs, clinical medicine formulated numerous useful diagnostic signs. Indeed, biophysical parameters such as elasticity and viscosity are some of the basic parameters of transport and mobility processes in a system. The value of the shear elasticity modulus of abnormal tissue is known to be up to 10–30 times that of normal tissue, whereas values of biochemical concentrations generally differ at most within 2–5 times ostensibly to preserve metabolic homeostasis. Hence, there is a need for an easily-affordable noninvasive methodology that could estimate elasticity and effusivity in tissue, especially for the brain, which is encased in bony enclosures, thereby providing considerable scope for using elastometric/effusiometric measurements to diagnose disease.

Keywords Biomarkers · Bioimaging · Non-invasive testing · Image-based diagnostics · Neurodegenerative diseases

1 Introduction

Historically the first biomarker to distinguish abnormal from normal tissue was biophysical properties, such as elasticity of tissue or viscosity of body fluids. Based on the stiffness or percussion of organs, clinical medicine formulated numerous useful diagnostic signs. Indeed, biophysical parameters such as elasticity and vis-

R. Ramaswamy (✉) · P. Roy
National Brain Research Centre, Manesar 122051, India
e-mail: rajiv@nbrc.ac.in

R. Khanna
Indian Institute of Technology, New Delhi 110015, India

R. Shorey and P. Ghosh (eds.), *Healthcare Engineering*,
DOI 10.1007/978-981-10-3111-3_11

cosity are some of the basic parameters of transport and mobility processes in a system. The value of the shear elasticity modulus of abnormal tissue is known to be up to 10–30 times that of normal tissue, whereas values of biochemical concentrations generally differ at most within 2–5 times ostensibly to preserve metabolic homeostasis. Hence, there is a need for an easily-affordable noninvasive methodology that could estimate elasticity and effusivity in tissue, especially for the brain, which is encased in bony enclosures, thereby providing considerable scope for using elastometric/effusiometric measurements to diagnose disease.

2 Methods

Using MRI-based image analysis, we develop a novel methodology to determine biophysical parameters without needing the customary procedure of piezoelectric mechanical vibration of brain tissue which might itself cause microtrauma in vascular disease patients. The elasticity modulus was determined by applying a numerical stress on each voxel of the three-dimensional MR image and calculating the deformation based on tissue elastic constants for white matter, grey matter and cerebrospinal fluid. Topological connectivity was determined as an imaging parameter using Minkowski functionals with the euler value representing connected voxels. We show that the two biophysical parameters of elasticity modulus and topological connectivity determined from T1-weighted magnetic resonance images are sufficient to discriminate normal controls, mild impairment and Alzheimer's disease with 96 % accuracy. The elasticity changes by about 80–120 % between normal and mild impairment, though level of biochemical (as amyloid-beta) in blood changes only by about 15 %.

3 Results

The resolution of the pre-AD MCI stage is nearly six times better with the use of biophysics-based markers such as elasticity modulus and topological connectivity as compared to biochemical markers which seem better suited to differentiating between diseased and non-diseased subjects. The determination of biophysical parameters also provides a more accurate diagnosis in neurodegenerative diseases where a marked structural change is seen as the disease progresses. The rates of decline can be used to diagnose and classify the disease in patients with varying levels of neurodegeneration.

4 Conclusion

Non-invasive neuroimaging biophysical markers are nearly six times more efficient at diagnosing ageing-related neurodegeneration and cognitive impairment than biochemical markers. The study also highlights the need for researchers and clinicians to shift focus to biophysics-related markers in the search for diagnostic and classification parameters of brain-related diseases. These parameters calculated from T1-weighted MR images also outperform other similar neuroimaging studies using fMRI, MRS, PET, DTI and DWI.

5 Discussion

Due to the recent development of biophysical instrumentation such as MRI/ultrasonography scanners, the diagnostic scope of biophysical markers, having high accuracy, needs to be proactively utilized and can be explored as biomarkers for other neurodegenerative diseases too.

References

1. Fiandaca MS, et al. The critical need for defining preclinical biomarkers in Alzheimer's disease. Alzheimer's & Dementia. 2014;10(3), (Suppl): S196–S212.
2. Wattamwar PR, et al. An overview of biomarkers in Alzheimer's disease. Annals of Indian Academy of Neurology. 2010;13, (Suppl2): S116–S123.
3. Riemenschneider MM, et al. Cerebrospinal Fluid Tau and β-Amyloid 42 Proteins Identify Alzheimer Disease in Subjects With Mild Cognitive Impairment. Archives of Neurology. 2002;59(11): 1729–1734.
4. Ramaswamy, R., R. Khanna, and P. K. Roy. Topological Connectivity and Elastic Responsivity of Brain as a window to the Alzheimer-type Neurodegenerative process. (in communication).
5. Mapstone M, et al. Nature Medicine. 2014;20: 415–418.
6. Roberts A.P. Morphology and thermal conductivity of model organic aerogels. Physical Review E. 1997;55: R1286–R1289.
7. Madadi M, et al. 3D Imaging and simulation of elastic properties of porous materials. Computing in Science and Engineering. 2009: 65–73.

Author Biography

Rajiv Ramaswamy Rajiv is a Senior Research Fellow in the Computational Neuroscience and Neuroimaging department of the National Brain Research Centre. His research involves the determination of physical and chemical properties of the human brain using structural MRI. These biophysical markers are used in differential diagnosis of patients with Alzheimer's Disease and early detection of the pre-AD state known as mild cognitive impairment. He has also been involved in fMRI work on language and cognition, and DTI-based white matter tractography to determine functional and structural connectivity in brain regions working in sync.

Rajiv has a BSc in Physics from Bangalore University and a Masters in Physics from the University of Sussex, England. He has also worked at the Siemens Corporate Technology Development Centre (Bangalore) and Manipal Hospital (Bangalore) before joining NBRC. Rajiv has presented posters at international conferences, has scientific publications to his name, and has recently filed a patent for his technique on AD detection and classification. His research interests lie in neuroimaging, biophysics, computer simulation and modeling.

Author Index

R. Shorey and P. Ghosh (eds.), *Healthcare Engineering*,
DOI 10.1007/978-981-10-3111-3

MIX
Papier aus verantwortungsvollen Quellen
Paper from responsible sources
FSC® C105338

If you have any concerns about our products,
you can contact us on
ProductSafety@springernature.com

In case Publisher is established outside the EU,
the EU authorized representative is:
Springer Nature Customer Service Center GmbH
Europaplatz 3, 69115 Heidelberg, Germany

Printed by Libri Plureos GmbH
in Hamburg, Germany